A First Course in "In Silico Medicine"

Volume 1

Series Editor
Masao Tanaka
Professor of Osaka University

1-3 Machikaneyama, Toyonaka
Osaka 560-8531, Japan
tanaka@me.es.osaka-u.ac.jp

Taishin Nomura · Yoshiyuki Asai

Harnessing Biological Complexity

An Introduction to Computational Physiology

 Springer

Taishin Nomura
Professor
Mechanical Science and Bioengineering
Graduate School of Engineering Science
Osaka University
1-3 Machikaneyama, Toyonaka
Osaka 560-8531, Japan
taishin@bpe.es.osaka-u.ac.jp

Yoshiyuki Asai
Specially Appointed Associate Professor
The Center for Advanced Medical
    Engineering and Informatics
Osaka University
1-3 Machikaneyama, Toyonaka
Osaka 560-8531, Japan
asai@bpe.es.osaka-u.ac.jp

ISBN 978-4-431-53879-0   e-ISBN 978-4-431-53880-6
DOI 10.1007/978-4-431-53880-6
Springer Tokyo Dordrecht Heidelberg London New York

Library of Congress Control Number: 2010940983

Springer is part of Springer Science+Business Media (www.springer.com)

Preface

The completion of human genome sequencing was an epoch-making event in reductionism biosciences, liberating a vast amount of experimental data and allowing us to relate specific parts of it to genes. The challenge for the biosciences in the twenty-first century is to integrate this information into a better understanding of biology, physiology, and human pathology. The integration is moving the world toward a new generation of bioscience and bioengineering, where biological, physiological, and pathological information from humans and other living animals can be quantitatively described *in silico* across multiple scales of time and size and through diverse hierarchies of organization - from molecules to cells and organs, to individuals. The physiome and systems biology represent such emerging biosciences. Such new trends in biosciences share a common direction, namely, an "integrative" approach that allows us to understand mechanisms underlying biological and physiological functions that emerge through the dynamics of each element and large aggregations of those elements.

The title of this textbook, *Harnessing Biological Complexity,* comes from the fact that we have a huge number of biological elements at hand - molecules, cells, and organs - and we are aware that each of them behaves in a more-than-simple way owing to its complex structure and the nonlinear dynamics governing its behavior. Moreover, an aggregation of specific elements behaves as a system. Thus, we are required to "harness" that complexity in terms of both the huge number of elements and their dynamics so that we can achieve an understanding of the mechanisms responsible for biological and physiological functions. Toward this end, integrative biosciences and bioengineering in their early stages aim at establishing frameworks and information infrastructures for describing biological structures and physiological functions on multiple scales of time and space. Next comes databasing and sharing, because it is not realistic to expect that a single research lab can find comprehensive answers to all the important questions. International communities such as those for systems biology markup language (SBML), CellML, JSim, Virtual Physiological Human (VPH), and the Neuroinformatics platforms of the International Neuroinformatics Coordinating Facility (INCF) have made pioneering efforts to establish those frameworks. In this textbook, we will use such a public platform, found at www.physiome.jp, which is being developed in communication with these other efforts mentioned above. The platform is capable of representing the hierarchical,

modular, and morphological structure of physiological entities and functional operations (influence) of one element on other elements. The platform provides a software
program called *insilico*IDE (ISIDE) which can take on the roles of a model com-
poser, a model browser, and a model simulator for models written in an XML format
such as *insilico*ML (ISML) which is used in physiome.jp, SBML, and CellML. The
platform has also a set of databases, *insilico*DB (ISDB), including ISML-ModelDB,
MorphologyDB, and Time-seriesDB. The platform can thus enhance the stream of
model sharing and increase the diversity of building blocks useful for integration.

Because biological and physiological functions are realized dynamically over
time, it is inevitable that we have to describe any one of them using mathematical
equations if we are to understand them quantitatively. We perform mathematical
modeling and computer simulations of a model for the function that is under consideration. Hence, this textbook is written as a brief introduction to computational
biology and physiology. Chapter 2 is on modeling at the cellular level, where we
consider several examples from electrophysiology. In Chap. 3, we deal with cellular
networks as models of tissues and organs. In both those chapters, we guide the reader
to grasp the physical and mathematical background of each of the models provided
and then to perform computer simulations on the models, to provide a deeper understanding of what is involved. The use of advanced physics and mathematics is
limited so that the textbook will be accessible to biologists and physiologists; illus-
trative details are provided when necessary. Nevertheless, it will be helpful for the
reader to have at least a basic knowledge of physics and mathematics as taught
at the first-year level of college. Simulations in the examples can be performed
easily because ISML files defining most of the models used in this textbook can
be downloaded from ModelDB at www.physiome.jp. Once *insilico*IDE has been
downloaded, the reader can just open a model file to simulate the dynamics of the
model. The reader can then alter parameter values and equations and construct her
or his own models to examine how they behave. The reader can also learn to use
the platform, including *insilico*IDE, in Chaps. 4 and 5, where there are more examples. After becoming familiar with the use of the platform, the reader can try
to construct a large-scale model for research purposes by combining the existing
models in the model databases at ModelDB of physiome.jp, the model repository
of CellML, and BioModels of SBML. In this way, new models can be created on
*insilico*IDE, and both existing and new models can be combined, meaning that we
are going to achieve a systematic way for harnessing biological complexity.

Chapters 1–3 were written by Taishin Nomura, and Chaps. 4 and 5 were written
by Yoshiyuki Asai. The authors express our sincere thanks to Dr. Hideki Oka,
Dr. Yoshiyuki Kido, Dr. Eric Heien, Dr. Yoshiyuki Kagiyama, Mr. Takahito Urai,
Mr. Tatsuhide Okamoto, Mr. Takeshi Abe, Ms. Li Li, Mr. Youichi Matsumura,
Dr. Ken Yano, Dr. Masao Okita, Dr. Rachid Ait-Haddou, and Dr. Daisuke
Yamasaki for their continuing efforts in developing and maintaining the plat-
form at www.physiome.jp. We also thank current and former students, especially
Mr. Yasuyuki Suzuki, Mr. Yousuke Yumikura, Mr. Masao Nakanishi, Mr. Toshihiro
Kawazu, and Mr. Shinsuke Odai, at the BioDynamics Laboratory of Osaka Uni-
versity for their research projects which contribute greatly to the development of

the platform. We are deeply grateful to Ms. Yukiko Ichihara at the BioDynamics Laboratory for the excellent illustrations in Chaps. 1–3. We very much appreciate the collaboration of Dr. Hiroaki Kitano, Dr. Yukiko Matsuoka, Dr. Samik Ghosh, and Dr. Akira Funahashi in the joint development of *insilico*IDE, making possible SBML–ISML hybrid simulations on the platform that are leading us in a new direction toward PhysioDesigner in the near future. We also thank Dr. Shiro Usui, Dr. Yoko Yamaguchi, Prof. Minoru Tsukada, Dr. Hiroaki Wagatsuma, and Dr. Yutaka Sakai for sharing database contents between INCF J-node platform and ModelDB at physiome.jp. The physiome.jp project is currently supported by the MEXT program, the Global Center of Excellence at Osaka University, chaired by Professor Yoshihisa Kurachi, Professor Ken-ichi Hagihara, and Taishin Nomura. Finally, we are grateful to Professor Masao Tanaka, the editor of this textbook series, for providing us with the opportunity to make our contribution to the series, and we thank the staff at Springer Japan for their patience and encouragement.

Taishin Nomura
Yoshiyuki Asai

Contents

Chapter 1
Introduction

A major part of biology has become a class of physical and mathematical sciences. We have started to feel, though still a little suspicious yet, that it will become possible to predict biological events that will happen in the future of one's life and to control some of them if desired so, based upon the understanding of genomic information of individuals and physical and chemical principles governing physiological functions of living organisms at multiple scale and level, from molecules to cells and organs.

Most efforts made in biology and bioscience researches today are still on a stage to identify or discover actors and actresses playing roles in every "story of a life." They are genes, molecules, cells, and organs. Each of them is related to specific biological and physiological functions. A number of functions and combinations of the functions are provided by different actors and actresses. An absence of a particular player changes the story, leading to disorder of a function or functions and causing an illness. Consequently biology and medical sciences should deal with complexity in terms of the huge number of actors and actresses. In order to handle this complexity, as one of varieties of solutions, biologists have been trying to establish databases[1] [2] [3] of genes and proteins for humans and different animals and plants. Medical researches require clinical databases at different levels of complexity[4] [5] to associate specific genes with their related diseases and also with dynamic time-varying signals that represent symptoms of the diseases. One of the ultimate goals of these efforts is to establish relationships logically and quantitatively based upon both clinical evidences stored in the databases as well as based upon underlying biophysical and physiological mechanisms that explain the relationships.

A single gene is a fragment of paired sequence of four bases known as adenine "A", cytosine "C", guanine "G", and thymine "T", embedded in a DNA (deoxyribonucleic acid) organized within chromosomes. A whole set of genes for

---

[1] http://www.ncbi.nlm.nih.gov/Genbank/.

[2] Examples are http://www.pantherdb.org/pathway/ and http://www.genome.jp/kegg/pathway.html.

[3] http://www.pdb.org/.

[4] https://cabig.nci.nih.gov/tools/c3d.

[5] http://physionet.cps.unizar.es/physiobank/database/mimic2cdb/.

T. Nomura and Y. Asai, *Harnessing Biological Complexity*, A First Course in "In Silico Medicine" 1, DOI 10.1007/978-4-431-53880-6_1, © Springer 2011

an individual is called *genome* which is a coined word connecting two words "gene" and "ome." The latter is a Latin word meaning "as a whole." The human genome project[6] has stated a complete of sequencing of a whole DNA of a human at year 2003. The human genome has about 3,000,000,000 base-pairs of ACGT, and may include about 20,000–25,000 genes.

A protein is a molecule that is made of amino acids chained by the peptide bonds, and it is synthesized in a cell using a gene that encodes a process for synthesizing the protein. Roughly speaking, there is a correspondence between a gene and a protein. However, a process for synthesizing a protein from its corresponding gene, called *gene expression* can interact with processes for synthesizing other proteins, introducing further complexity in terms of the variety of the proteins and their synthesizing processes. More precisely, we need another type of actors for the protein synthesis. They are RNA (ribonucleic acids) which are created or transcribed from DNA as an equivalent copy of a sequence of DNA using enzymes. The processes are called *transcription*. Revealing a whole set of RNA or transcripts for an organism is called *transcriptome*, coined for meaning transcriptions as a whole. Thus, the road to identifying all correspondences between the genes and proteins is still largely in progress. For example, UniProtKB/Swiss-Prot[7] provides a database of protein sequences representing the order of amino acid residues chained by peptide bonds and information relating genes and their corresponding proteins. They also describe the function of every protein and its domains structure. Revealing all of these mappings or a entire of set of proteins is called *proteome*, coined for meaning proteins as a whole.

A process of protein synthesis is governed by a sequence of bio-chemical reactions of molecules, and thus it is a dynamic process. Again, a number of molecules distributed in a cell can be involved in this process. A sequence of reactions of the molecules involved is referred to as a *pathway*, regulating the corresponding gene expression (*gene regulation*). Note that such sequences of processes for a protein synthesis can also be complicated not only because of the number of the players but also because chemical products of every reaction along the sequence can affect reactions of upstream and downstream of the sequence in manner of feedback control. The research field called *systems biology* deals with such complicated pathways, in which mathematical modeling of sequences of reactions with the use of related biological databases is studied to reveal emerging properties of the reactions. In order to overcome the complexity in terms of pathway network as well as the numbers of pathways, public efforts have been promoted, in which pathway databases (see footnote 2) and databases of their mathematical models[8] have been established. SBML,[9] SBGN,[10] and their related tools are now widely used in the community of the systems biology.

---

[6] http://www.genome.gov/10001772.

[7] http://www.uniprot.org/.

[8] http://www.ebi.ac.uk/biomodels-main/.

[9] http://sbml.org/.

[10] http://sbgn.org/.

A protein molecule is a set of atoms and a set of carboxyl and amino groups interacting electromagnetically. This is a world of quantum physics and chemistry, a major field of *biophysics*. Primary important information on a protein is its peptide sequence as mentioned above (see footnotes 2 and 7). A three-dimensional configuration of atoms for a given protein is called *protein structure*, and a set of solved structure is databased in Protein Data Bank (see footnote 3).[11] Conformational changes in a protein provide a function or functions of the protein. Thus, one can say that biological functions emerge through "dynamics." Non-dynamic, statically "freezed" proteins are just materials and may not be able to perform any functions. The muscle protein called myosin is a typical example of the motor proteins that plays an important role in contraction of muscle fibers. Conformation changes in the myosins make them walk along actin filaments, other motor proteins. Energy necessary for those structure changes are supplied by the hydrolysis of ATP, by which chemical energy is converted into the mechanical work. A major methodology for quantitative description of dynamic changes in a protein structure is the molecular dynamics or MD simulation which is usually based on classical equations of motion as well as quantum physics.

Cells are the basic building block of a life of organism. They involve a number of proteins maintaining cellular functions such as cell metabolisms and other electrochemical activities to maintain themselves, to perform their functions, and to communicate with other cells. Physiological tissues and organs are made of the cells, which perform functions as cellular network systems. Characterization of a cell or a cell type can be achieved in varied ways. For examples, one way is to use proteome-based methodologies for analyzing protein compositions of organelles and cells.[12] Similarly, cell signaling pathways (see footnote 2)[13] is used to characterize cells. Cell morphology (geometry)[14] is also another way to classify and understand cellular functions. These different types of information are integrated in a cell to determine how the cell performs its functions.

A brief story described here is a bottom-up view of biological and physiological functions from genes and proteins to cells and organs. The new emergent research field called *physiome* aims to integrate these pieces of information at different scales. The physiome is also coined for representing "physio," meaning a life, as a whole. Ambition of the physiome is to establish quantitative bridges among those biological structure and functions and to integrate biological and physiological functions at multiple levels of time and space, through databasing, modeling, and

---

[11] http://www.pdbj.org/.

[12] http://www.mapuproteome.com/.

[13] http://stke.sciencemag.org/cm/.

[14] http://ccdb.ucsd.edu/.

simulating the functions. International communities have been making efforts for this purpose[15] [16] [17] [18] (Bassingthwaighte 2000; Hunter and Borg 2003).

Figure 1.1 illustrates a hierarchical or physiomic view of a skeletal muscle, from a molecular level of myosin and actin proteins (10–100 nm scale), sarcomere, a myofibril, bundle fibers (a muscle fiber, which is a long cylindrical multi-nucleated cell), up to the one muscle group (0.1 m scale) as an organ level. The macroscopic function of the muscle group in generating a contraction force f may be modeled simply based on its spring-dashpot-like property as shown in Fig. 1.2, in which the active contractile element (CE), the passive elastic spring element with its spring

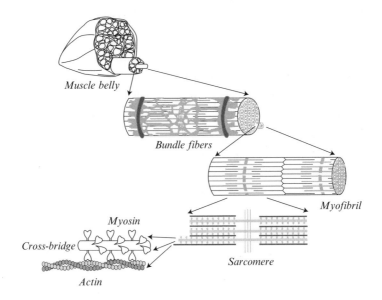

Fig. 1.1 A muscle at multiple scales and levels

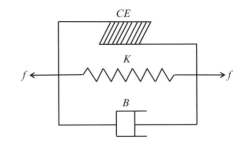

Fig. 1.2 A macroscopic
mechanical model of a
muscle

[15] http://www.physiome.org.nz/.

[16] http://www.vph-noe.eu/ and http://www.biomedtown.org/.

[17] http://www.physiome.org/.

[18] http://www.physiome.jp/.

constant K, and the passive viscous element with its viscosity B are arranged in parallel (McMahon 1984). The spring-dashpot-like property is modulated by muscle activity which is determined or controlled by the calcium influx in response to neural stimulations delivered from the central nervous system (motor neurons) to the muscle fiber via a neuromuscular junction, the point where a nerve axon from the nervous system and the muscle fiber contact. Let us briefly consider a model of the passive elastic spring element. It is often simply modeled as a linear spring with its spring constant K, meaning that it obeys the Hooke's law, i.e., it exerts a restoring force Kx if the element is elongated by the amount of x from its equilibrium (resting) length. Experiments have shown that the muscle and muscle fiber obey this law at least for small elongations (McMahon 1984). How this experimental rule can be associated with the muscle and muscle fiber as the physical and biological entities? It has been shown that the biological origin of the passive elasticity of muscles is the largest known protein called titin or connectin (Maruyama et al. 1984) which is a part of the sarcomere. It is considered that a part of the spring-like elasticity is "entropic" as in a rubber. The statistical physics tells us about the entropic elasticity as follows. A chain made of a number of elements (say $n \gg 1$) freely joined and folded in one-dimensional straight line space, which is a very simple model of a single connectin protein or an in situ muscle fiber made of a set of sarcomeres, can take a number of configurations for a given total length x. Let us denote this number of configuration as $W(x)$ as a function of x. The more variety of configurations can the chain take for a given x, the larger is the number $W(x)$. Let us define the *entropy* of the system as

$$S(x) = k_B \ln W(x), \tag{1.1}$$

where k_B is the Boltzmann constant, as formulated by Ludwig Eduard Boltzmann (Fig. 1.3). The configuration number $W(x)$ determines the entropy $S(x)$ of the chain system. The larger the number of $W(x)$ is, the larger is the value of the entropy $S(x)$. Let X be a force required to keep the length of the chain at x. This is a "statistical" or "phenomenological" force induced by a statistical tendency to move a state of the chain toward more stable state, the thermodynamic equilibrium, so that the system's state changes as the free energy defined as $F = E - TS$ of the chain system with its temperature T decreases, where E represents the internal energy of the system. If we consider a case where the temperature of the chain system is constant at T, the force is determined only by changes in the entropy of the system. This is because the temperature of the chain is constant, and there is no dissipation of the internal energy E from the system with the friction free joints, and also because there is no gain of the internal energy into the system from outer environment, meaning that the internal energy of the system does not change even if the chain configuration changes. The force thus describes a statistical tendency for the chain system to increase its entropy, by which the free energy F decreases. The force X is defined as the free energy gradient with respect to x, and it is obtained as

$$X = \left(\frac{\partial F}{\partial x} \right)_T = -T \left(\frac{\partial S(x)}{\partial x} \right)_T \tag{1.2}$$

where the subscript T means that the value of the derivative is evaluated at the given value of T.

Fig. 1.3 $S = k_B \ln W$ on the
gravestone of Ludwig Eduard
Boltzmann

Fig. 1.3 $S = k_B \ln W$ on the gravestone of Ludwig Eduard Boltzmann

Let n^+ and n^- be the numbers of elements in positive and negative direction, respectively, along the coordinate of the one-dimensional chain line with $n = n^+ + n^-$. Assuming all elements are identical and the length of each element is a, n^+ and n^- are related to the length of the chain x as $x = (n^+ - n^-)a$. Then, the number of configuration $W(x)$ that gives rise to the chain of the length x is given by the combinatorial number that one can take any n^+ elements from the n elements as

$$W(x) = \frac{n!}{n^+!n^-!}. \tag{1.3}$$

Since the entropy is a monotonically increasing function of the configuration number, the larger the number of $W(x)$, the more stable is the state of the chain in terms of the entropy. Thus, the entropic force is generated so that the chain tends to be shorter. Substituting (1.3) into (1.1) and then (1.2), and assuming further that $x \ll na$, i.e., the chain elements are multiply folded, we have

$$X \sim \frac{k_B T}{2a} \ln \frac{1 + x/(na)}{1 - x/(na)} = \frac{k_B T}{na^2} x + \cdots . \tag{1.4}$$

This shows that, by considering only linear term of x,

$$\frac{k_B T}{na^2} \tag{1.5}$$

represents the spring constant of the chain for a short length x. Although the chain considered here is a too much simplified model of the muscle fiber, this exemplifies a possible approach that can be used for relating protein configurations at the molecular level to a macroscopic concept, the passive spring constant of the muscle fiber in this case. Recent studies are going to uncover more precise relationships between the molecular structure of the titin (connectin) and the elasticity of the titin and the muscle fiber (Li et al. 2002; Lee et al. 2007), even for cases that the elongation x is large. Statistical physics aims at relating microscopic properties of molecules to macroscopic characteristics of ensemble of the molecules. In the example here, the mechanical spring property of the muscle is roughly related to characteristics of the proteins. Establishing such relationships might be possible based on classic statistical physics if we consider relatively non-dynamic situations in which systems are close to their thermodynamic equilibria or if we can assume adiabatic or quasi-adiabatic processes. "Passive" properties, as the passive elasticity considered here, may be in such conditions.

Exercise 1.1. Derive (1.4) using Stirling's approximation for a large number N

$$\ln N! \sim N \ln N - N,$$

and a Taylor series expansion of the function $f(\epsilon) = (1 + \epsilon)/(1 - \epsilon)$.

Studies for bridging between active contraction properties in the macroscopic CE element in Fig. 1.2 and that for an actomyosin filament at the molecular level as illustrated in Fig. 1.1 are actively promoted today (Reimann 2002). The macroscopic CE element can be characterized by its nonlinear force–length and force–velocity relationships (Fig. 1.4), referred to typically as the Hill-type model after the famous physiologist (Hill 1938).

In a microscopic view, many active agents are in charge of the muscle contraction. Their behaviors are no more linear. Moreover, energies are generated and also consumed by the agents. These make it difficult to establish a bridge between microscopic properties of motor molecules and macroscopic characteristics of the CE element as the ensemble of the molecules. For a contraction of an active contractile element, influx of calcium ions Ca^{2+} into the muscle fiber is required, and it is usually triggered by signals from the central nervous system, inducing a sequence of electrical and chemical processes in the muscle fiber membrane. In Chap. 2, we will look at such activities. In those processes, chemical

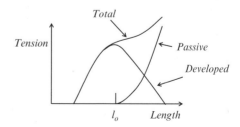

Fig. 1.4 A force–length relation of a muscle as a sum of passive and active (developed) tension

energies are used to alter molecular configuration of the myosin proteins, by which they are converted to mechanical energies to move (walk) the myosins along the actin filaments. Energies are also consumed for restoring the Ca^{2+} ions used in the contraction as well as pumping them out the membrane.

There have been a top-down view of physiological functions in the classic research field called *biocybernetics* or more contemporary *systems physiology*, in which populations of individuals, organs, and cells are considered as "dynamical systems" with multiple feedback loops of signal transductions and communications. It seems that the ideas developed by the classical cybernetics are inherited in most of the contemporary "omics" biosciences, although the cybernetics tends to describe physiological functions as simple as possible in order to elucidate and understand essential mechanisms of the functions, rather than describing them in detail as large as life, which might be the basic objective of the "omics" biosciences.

Mathematics plays an important role for describing biological and physiological functions. This is because, as mentioned above, these functions emerge through "dynamics." That is, any biological and physiological function emerges though time-dependent changes in "a state" of a system that is composed of biological materials. "A state" is not a material, but just a man-made concept. We need to realize that any principle of physics formulates how a state of physical material changes in time. We expect it is the same in biosciences. We need to realize that considering states of biological entities is equally important as discovering genes and protein structure. Once a biological function is formulated as a set of mathematical equations, i.e., a mathematical model of the function to describe how a state of a biological entity changes in time, we may be able to understand how the function emerges and to predict events that may happen in the future for a given observation of its initial state. If the prediction does not well match its experimental validation, the modeling and thus our understanding, are wrong, and we need to improve the model and our understanding. Mathematical models can be analyzed in detail via computer simulations as well as analytical methodologies. As a state of gas molecules can be described by both a set of equations of the statistical physics and a set of motion equations of molecular dynamics, biological functions might be described by different methodologies at different levels. However, there might be logical and theoretical relationships between the levels. This text book introduces bases of mathematical modeling and analyses of biological and physiological phenomena with keeping this hope in mind.

Among huge amount of existing models, we mainly look at electrophysiological models of cellular dynamics and networks of interacting cells through examples in this textbook. For simulating dynamics of the models, we use a software platform called *in silico* platform developed by public effort at www.physiome.jp. The platform includes model databases (ModelDB, MorphologyDB, and TimeseriesDB), and an integrated environment called *insilico*IDE as a simulator. The *insilico*IDE can also be used as a model browser and a model editor. Most of the examples used in this textbook can be downloaded from the ModelDB at www.physiome.jp. The user can examine dynamics of each model example by executing it. This is one of the novel aspects of this textbook. Since the *insilico*IDE can simulate models written in the languages developed by other platforms such as SBML and CellML, the user might be able to construct own models by combining different types of models at different levels through experiences on this textbook. It is still very open problem and challenge to establish better methodologies that can bridge between different levels of biological and physiological functions.

Chapter 2
Modeling Cellular Dynamics

Cells are the basic units of biological structure and functions. They make up tissues and our bodies. A single cell includes organelles and intracellular solutions, and it is separated from outer environment of extracellular liquid surrounding the cell by its cell membrane (plasma membrane), generating differences in concentrations of ions and molecules including enzymes. The differences in charges of ions and concentrations cause, respectively, electrical and chemical potentials, generating transportations of materials across the membrane. Here we look at cores of mathematical modeling associated with dynamic behaviors of single cells as well as bases of numerical simulations.

2.1 Ion Channels and Cellular Excitations

The difference in the intra and extra-cellular electrical potential is referred to as the *membrane potential*. The cellular membrane potential shows dynamic changes either spontaneously or in response to external stimulations that are electrical, chemical, and/or mechanical.

Figure 2.1 exemplifies such changes in the membrane potential in neuronal cells of a cat spinal cord, in which the membrane potential shows a single spiky change on the left (A) and repetitive burst-like changes on the right (B). Each of the spiky transient increase and decrease is referred to as the *action potential*, or in cases of neuronal cells, the *spike*. In Fig. 2.1 (B), in particular for the set of two traces at the middle, we can observe two different phases alternate almost periodically. In one phase, spikes are generated with high frequency (active phase of the burst), and no spikes in the other phase (silent phase of the burst). Action potentials play significantly important roles for functions of cells, and thus for functions of cellular tissues and organs. The neuronal cells used in the experiment for Fig. 2.1 are associated with the control of muscle contractions actuating motions of the animal's limbs during rhythmic walking. This means that the sequences of action potentials of the neuronal cells "encode" and transmit information on the muscular force intensities and timings, i.e., how much the muscles that are controlled somehow by those cells would generate the force and how the force would change along the time.

T. Nomura and Y. Asai, *Harnessing Biological Complexity*, A First Course in "In Silico Medicine" 1, DOI 10.1007/978-4-431-53880-6_2, © Springer 2011

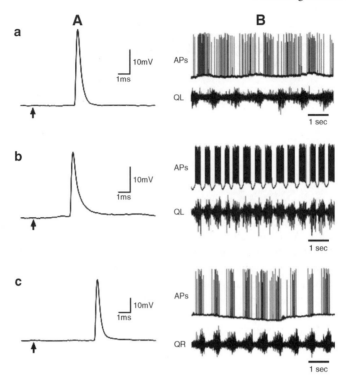

Fig. 2.1 (**A**) Action potentials of spinal commissural interneurons (**a, b**) and a spinal neuron with a long ascending axon (**c**), all of which were recorded intraaxonally in the lumbar spinal cord of decerebrate, immobilized cats. The potentials in **a** and **b** were evoked monosynaptically by stimulation to the medullary reticular formation. The potential in **c** were evoked antidromically by stimulation to the same reticular area, thus indicating that the neuron extend its axon rostrally up to the reticular formation in the brainstem. Upward arrows in **a–c** indicate timings of the stimulation. (**B**) Locomotion-related discharge activities of three spinal neurons **a–c** during the generation of fictive locomotion, which was evoked by stimulation to the mesencephalic locomotor region. In each figure, the upper trace represents rhythmic discharges of neuronal action potentials (APs), and the lower trace shows activities of electroneurograms recorded from the left or right quadriceps nerves (QL or QR). This figure is made available here through the courtesy of Prof. Kiyoji Matsuyama

Indeed, this can be better understood by looking at the lower trace of each of three recordings (upper, middle, and lower) of Fig. 2.1B, which is the electroneurogram recorded from the left or right quadriceps nerves, representing the total neural command delivered to the muscle to induce muscle contractions. In the middle and lower traces of Fig. 2.1B, the active phases of the burst correspond to the large activities of the electroneurogram, implying that the neural system commands the muscle to perform large contractions.

Muscle cells such as cardiac heart cells and skeletal muscles also generate action potentials that control contractions of the muscle fibres as we will see later in this chapter. Pancreatic β cells show burst-like action potentials in response to an increase in glucose concentration in the blood in order to control the amount of insulin gene expression and secretion as we will also see briefly in this chapter.

Fig. 2.2 Schematic
illustration of cross-section
of a single ion channel
protein that forms a pore and
is embedded in the
lipid-bilayer of the cellular
membrane

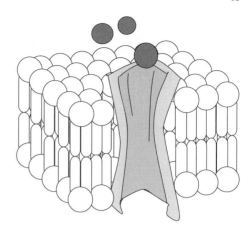

Action potentials and many other dynamic activities of a cell are induced and/or modulated by communications between the cell and its environment. Such communications are achieved through membrane channels and receptors. Figure 2.2 schematically illustrates an ion channel protein which is embedded in the membrane to allow flows of ions across the membrane. Such a protein forms a pore and controls influx and efflux of several kinds of ions. The flow of ions can generate dynamic changes in the membrane potential which plays important roles for intra-cellular signal transmission such as calcium signaling as well as inter-cellular communications at tissue level.

2.1.1 A Simple Open–Close Kinetics of a Channel Protein

A typical ion channel allows influx and efflux of a specific type of ions. This property is referred to as the *ion selectivity*. For example, a single potassium channel protein forms a pore which is selective to potassium ions K^+. Ion channels are not merely passive pores, but they actively controls the flow of ions by opening and closing the channels. Let us consider a large number of a specific type of channels embedded in a membrane of a single cell. Let N be the total number of the channels. They allow the flow of a specific type of ions. We assume, as the simplest case, that each of the channels takes one of two states, namely "closed" and "open." See Fig. 2.3. The open state allows the ions to flow through the channel, but the closed state does not. Let $p(t)$ be a fraction of the open state over all the channels at a time instant t, i.e.,

$$p(t) = \frac{\text{\# of the channels at the open state}}{N}$$

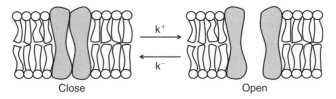

Fig. 2.3 A simple open–close kinetics for a single ionic channel with the rate constants k^- and k^+

Then the fraction of the closed state $q(t)$ can be expressed as

$$q(t) = \frac{\text{\# of the channels at the close state}}{N}$$

$$= \frac{N - \text{\# of the channels at the open state}}{N}$$

$$= 1 - p(t)$$

Let us consider a single channel among $Np(t)$ at the open state at time t. The channel makes a transition from the open to the closed state stochastically. If the transition is governed by a Poisson point process with its rate or intensity k^-, the probability that a single transition occurs for an infinitesimal small time interval Δt is $k^- \Delta t$. Indeed, this is a definition of the Poisson point process. Moreover if every channel at the open state makes such a transition independently, the mean (ensemble averaged through all the channels) number of the transitions from the open to the closed state within Δt would be $Np(t) \cdot k^- \Delta t$. The number k^- is referred to as the *rate constant*. The reverse transitions can be described similarly by considering the probability that a single transition from the closed state to the open state occurs for an infinitesimal small time interval Δt is $k^+ \Delta t$. Thus, the mean number of channels at the open state at time $t + \Delta t$ can be described as follows:

$$Np(t + \Delta t) = Np(t) - Np(t) \cdot k^- \Delta t + N\,(1 - p(t)) \cdot k^+ \Delta t$$

which is reformulated as

$$\frac{p(t + \Delta t) - p(t)}{\Delta t} = -k^- p(t) + k^+\,(1 - p(t))$$

It becomes an ordinary differential equation (ODE) as $\Delta t \to 0$, i.e.,

$$\frac{dp(t)}{dt} = -k^- p(t) + k^+\,(1 - p(t))$$

$$= -(k^- + k^+)\left(p(t) - \frac{k^+}{k^- + k^+}\right). \tag{2.1}$$

Defining $\tau = 1/(k^- + k^+)$ and $p_\infty = k^+/(k^- + k^+)$, we have

$$\frac{dp}{dt} = -\frac{p - p_\infty}{\tau} \tag{2.2}$$

A solution of (2.2) can be obtained analytically by integrating (2.2). This can be done as follows. First, put $x(t) = p(t) - p_\infty$ for the ease of our calculation. Then $dx(t)/dt = dp(t)/dt$, i.e., $dx = dp$, since p_∞ is constant, leading to a new ODE for x as

$$\frac{dx}{dt} = -\frac{x}{\tau}$$

with $x(0) = p(0) - p_\infty$, the initial condition, where $p(0)$ is the open fraction of the channel at the initial time $t = 0$. This can be further rewritten as

$$\frac{dx}{x} = -\frac{1}{\tau}dt.$$

Integrating both sides by x and u (representing time here instead of t), we have

$$\int_{x(0)}^{x(t)} \frac{dx}{x} = -\int_0^t \frac{1}{\tau}du, \tag{2.3}$$

obtaining the solution of (2.2) in terms of $p(t)$ as

$$p(t) = p(0)\exp\left(-\frac{t}{\tau}\right) + p_\infty\left(1 - \exp\left(-\frac{t}{\tau}\right)\right) \tag{2.4}$$

Note that the dynamic variable $p(t)$ of the ODE represents either the open fraction of the channel at time t if we consider all N channels on the membrane or the open probability if we consider a single channel among N channels.

Exercise 2.1. Derive (2.2) from (2.1). Perform the integration in (2.3) to find $x(t)$. Then transform $x(t)$ back to $p(t)$ to obtain the solution as in (2.4).

IDE Modeling: A Simple Channel Kinetics: Search ModelDB by "Kinetics" to Find Simple_Channel_Model.isml

$$\frac{dp}{dt} = -\frac{p - p_\infty}{\tau} \tag{2.5}$$

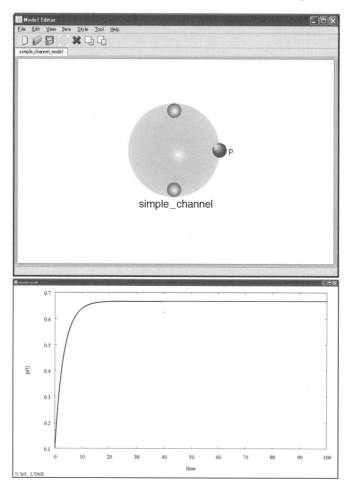

Fig. 2.4 IDE modeling of a simple channel kinetics. *Upper*: Screen shot of the *insilico*IDE platform for the model which is represented as the ball-like module on the canvas. *Lower*: Dynamics of a simple channel model of (2.2) simulated using *insilico*IDE platform. An initial condition is set as $p(0) = 0.1$. The parameter values used here are $k^+ = 0.2$ and $k^{-1} = 0.1$

Figure 2.4 shows a screen snapshot of this model at the upper panel and a numerical solution of (2.2) or (2.5) at the lower panel. We use a tool called the *insilico*IDE of *insilico* platform (see footnote 18). Hereafter in this textbook, we will frequently use this platform provided by *Physiome.jp* at http://www.physiome.jp. It is one of the public efforts for promoting sharing of mathematical models of physiological functions and their related information. It is on the public domain and can be downloaded for free and used for simulating and/or constructing models of physiological functions. The platform also provides a standard format for describing mathematical models of physiological functions. It is based on a XML technology, and the format is called *insilico*ML (*insilico* Markup Language) abbreviated as ISML. A single ISML file includes all information necessary for a numerical simulation of the model such as equations of the model, text explanation of the model as well as published articles related to the model. See later part of this textbook for details of the ISML format and the *insilico*IDE. In the upper panel of the figure, a ball-like object is displayed on a white canvas of the *insilico*IDE. We call such a ball-like object the *module*, representing a specific physiological function or a set of functions. In this example, the module represents the membrane channel kinetics modeled by (2.5). The module has a site from which the user can obtain internal information of the module called the *out-port* and it is represented as a small circle located on the right edge of the ball-like module. In this example, a value of the state variable $p(t)$ of the model can be obtained through the output port. It is a solution of the model when we use the *insilico*IDE as a model simulator. The lower panel of the figure is obtained in such a way for an initial condition $p(0) = 0.1$ specified in the module. The parameter values of the model are also defined in the module, and in this case, they are $k^+ = 0.2$ and $k^- = 0.1$. The numerical solution asymptotes to 0.66 which is identical to $p_\infty = k^+/(k^- + k^+) = 0.2/(0.2 + 0.1) \sim 0.66$ as predicted by the analytical expression in (2.4). It can be observed from the time course of the solution that the *time constant* of this simple dynamics is $\tau = 1/(k^- + k^+) = 1/(0.2 + 0.1) \sim 3.3$ s. That is, starting from $p(0) = 0.1$ at $t = 0$ s, $e^{-1} \times 100\%$ of the final asymptotic value 0.66 is attained at the time $t = \tau \sim 3.3$ s. The smaller the time constant is, the faster are the dynamics and the shorter is this time interval.

2.2 First Encounter with Numerical Simulations of Ordinary Differential Equations

Equation (2.2) can be discretized as follow.

$$\frac{p(t + \Delta t) - p(t)}{\Delta t} = -\frac{p(t) - p_\infty}{\tau} \equiv f(p(t))$$

Rearranging the discretized equation, we have

$$p(t + \Delta t) = p(t) + \Delta t\, f(p(t))$$

For a given initial condition $p(0)$ at $t = 0$, one can obtain the state variable $p(\Delta t)$ at $t = \Delta t$ as

$$p(\Delta t) = p(0) + \Delta t\, f(p(0)).$$

We repeat the same to have the values of p at $t = 2\Delta t, \ldots, n\Delta t$ as follows:

$$p(2\Delta t) = p(\Delta t) + \Delta t f(p(\Delta t))$$

$$\vdots$$

$$p(n\Delta t) = p\left((n-1)\Delta t\right) + \Delta t f(p((n-1)\Delta t))$$

This type of simple numerical integration is called the *forward Euler discretization scheme*. Here we consider a very simple ODE

$$\frac{dp}{dt} = -\frac{p}{\tau} \tag{2.6}$$

by setting $p_\infty = 0$ in (2.5) or by the use of the variable transform as $p \to p - p_\infty$ to apply the forward Euler scheme. The iterative applications of the Δt time evolution for a given initial condition $p(0)$ yields

$$p(n\Delta t) = p((n-1)\Delta t) - \frac{\Delta t}{\tau} p((n-1)\Delta t))$$

$$= \left(1 - \frac{\Delta t}{\tau}\right) p((n-1)\Delta t)$$

$$= \left(1 - \frac{\Delta t}{\tau}\right)^n p(0)$$

$p(n\Delta t)$ must be converging to zero as $n \to \infty$ for a positive τ, and it is the case if we take Δt smaller than τ. However $p(n\Delta t)$ will show a damped oscillation or oscillatory explosion if we take a large step of time discretization for Δt which makes $(1 - \Delta t/\tau) < 0$, implying a numerical instability.

The derivative of $p(t)$ in (2.2) can also be discretized differently to have the following approximated equation.

$$\frac{p(t) - p(t - \Delta t)}{\Delta t} = f(p(t)).$$

One should avoid a pitfall of taking $p(t - \Delta t)$ for the argument of the function f at the right-hand-side of (2.2). That is, one should note that only $dp(t)/dt$ is approximated differently from the case above. Then we have

$$p(t) = p(t - \Delta t) + \Delta t f(p(t)). \tag{2.7}$$

This is called the *backward (implicit) Euler discretization scheme* to obtain the state variable $p(t)$ at time t from a given $p(t - \Delta t)$ at time $t - \Delta t$, one time step past. In order to obtain the unknown variable $p(t)$ from a known value of $p(t - \Delta t)$, we need to solve (2.7) with respect to $p(t)$. To this end, (2.7) is rearranged as

$$p(t) - \Delta t f(p(t)) = p(t - \Delta t).$$

Let us define a function $G(p(t))$ as

$$G(p(t)) \equiv p(t) - \Delta t f(p(t))$$

to have the algebraic equation

$$G(p(t)) = p(t - \Delta t).$$

Then $p(t)$ can be obtained by solving formally as

$$p(t) = G^{-1}(p(t - \Delta t))$$

if G is locally invertible around $p(t)$. Equivalently, $p(t)$ can be obtained as a solution (zero) of the algebraic equation $G(p(t)) - p(t - \Delta t) = 0$.

For the simple example of (2.6) used above,

$$G(p(t)) \equiv \left(1 + \frac{\Delta t}{\tau}\right) p(t)$$

and we need to solve

$$G(p(t)) = p(t - \Delta t)$$

yields

$$p(t) = \frac{1}{\left(1 + \frac{\Delta t}{\tau}\right)} p(t - \Delta t).$$

Thus

$$p(n \Delta t) = \frac{1}{\left(1 + \frac{\Delta t}{\tau}\right)^n} p(0)$$

which is always converging to zero regardless of the value Δt. This means that the implicit Euler is more robust than the forward Euler in terms of the numerical stability.

2.3 Gating of Ion Channels in the Hodgkin–Huxley Formulation

In 1952, Alan Hodgkin and Andrew Huxley published a paper that summarized essential mechanisms explaining dynamic electrical behaviors of cellular membranes referred to as the *membrane excitation*, that is, to describe the action potential, using squid giant axons (Hodgkin and Huxley 1952). They received the 1963 Nobel Prize in Physiology and Medicine for this work. Their formulation is well-known as the *Hodgkin–Huxley model* (HH model) or the HH equation.

Figure 2.5 illustrates their modeling concept, in which three kinds of ion channels embedded in the axonal membrane are considered. They are the potassium ion channel that is specifically permeable to K^+ ions, the sodium ion channel that is specifically permeable to Na^+ ions, and the chloride ion channel or so called leak channel that is permeable to Cl^- ions.

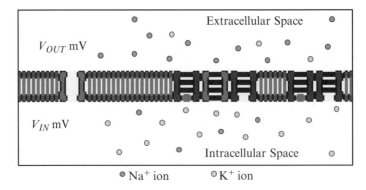

Fig. 2.5 A schematic illustration of ion channel kinetics in the Hodgkin–Huxley formulation

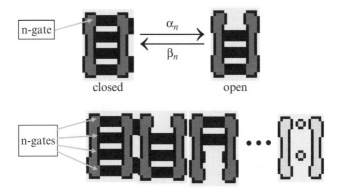

Fig. 2.6 Gating of the potassium channel in the Hodgkin–Huxley formulation

The basic idea of the HH model is the same as the simple channel that we have analyzed for (2.2). However, the open–close kinetics is slightly complicated. Hodgkin and Huxley proposed the concept of *gating* which can describe the opening (activation) or closing (inactivation) of ion channels. In terms of gating, a single channel in (2.2) has a single gate whose opening and closing correspond to the channel opening and closing, respectively. In general, a single ion channel can have several gates. In this case, an open state of the channel is achieved when all of the gates associated with the channel are in their open states.

Let us begin with modeling of the *potassium ion channel*. Hodgkin and Huxley hypothesized that a single potassium channel possesses four identical gates, each of which is referred to as the n-gate. See Fig. 2.6 visualizing this situation in which a cartoon of the single K^+ channel at the top-left corner has four horizontal bars representing four n-gates associated with this channel. The upper part of Fig. 2.6 shows a state transition of the most upper n-gate of the channel where the gate is closed on the left, and it is open on the right. The rate constant for the closed to open transition is α_n, and that for the open to closed is β_n. The HH model assumes the identicality and stochastically independence of four n-gate kinetics for each of

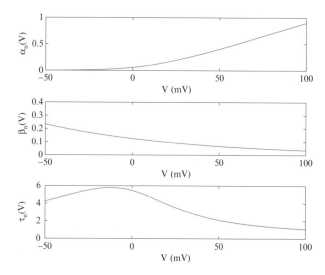

Fig. 2.7 Membrane potential dependency of $\alpha_n(V)$, $\beta_n(V)$ and $\tau_n(V)$ of the n-gate kinetics of the potassium channel

the K^+ channel. Because of the stochastic independence of the gates, opening and closing of one of four gates do not affect that of the other gates. The lower part of Fig. 2.6 shows several different states of the potassium channel. Three situations on the left are all representing closed states where all four gates are closed in the leftmost case, the only top gate is open with the other three gates closed in the middle case, and the top and the second gates are closed with the third and the bottom gates open in the right case. There are several combinations for such open–closed gates. The case in which all of four gates are open as illustrated in the right-most cartoon of the lower part of Fig. 2.6 represents the open state of the potassium channel, allowing influx and efflux of potassium ions.

The HH model assumes that the rate "constants" α_n and β_n are not constant, but they change as the functions of the membrane potential V which is the difference in the electrical potential inside and outside the membrane as we will define later in this chapter. As shown in Fig. 2.7, $\alpha_n(V)$ and $\beta_n(V)$ are, respectively, monotonically increasing and decreasing functions of V. Because of this, when the membrane potential V is low, β_n is greater than α_n, allowing more frequent transitions of n-gates from open to closed. Thus the channel is more likely in the closed state. When V is high, α_n is greater than β_n, and it allows n-gates to make more frequent transitions from closed to open. Thus the channel is more likely in the open state.

IDE Modeling: n-Gate Model of the HH Potassium Channel: Search ModelDB by "n Gate" to Find n_Gate_Model.isml

$$\frac{dn}{dt} = -\frac{n - n_\infty(V_{clamp})}{\tau_n(V_{clamp})} \tag{2.8}$$

Equation (2.8) represents the n-gate open–close kinetics. Hodgkin and Huxley identified the dynamics of n-gate using an electrophysiological techniques called the *voltage clamp* in which the membrane potential, i.e., the electrical potential difference between inside and outside of the membrane is clamped (fixed). They observed the potassium ion current in response to a step-wise constant clamp voltage. Here we simulate (2.8) to examine dynamic changes in the opening probability of the n-gate in response to a step-wise constant clamp voltage (V_{clamp}). Figure 2.8 shows a modeling of the voltage clamp *in silico* experiment (upper panel) and a numerical solution for (2.8) in response to a step-wise voltage clamp (lower panel). In the upper panel, we have two modules, representing specific functions. The left module represents a voltage step generator and the right module a single n-gate.

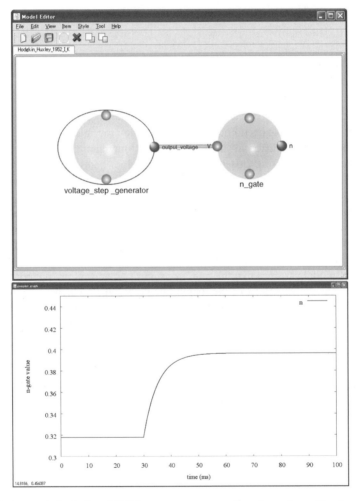

Fig. 2.8 A numerical solution of (2.8) in response to a step-wise voltage clamp. The clamp voltage for the first 30 ms is 0 mV and that after 30 ms is 5 mV. The initial value $n(0)$ is set to $n_\infty(0) \sim 0.32$

The step voltage generated by the left module is delivered to the right module through the line connecting these two modules. The clamp voltage for the first 30 ms is set to 0 mV, and then it becomes $V_{clamp} = 5$ mV after $t = 30$ ms as these parameters are defined in the module of the voltage step generator. The initial value $n(0)$ is set to $n_\infty(0) = \alpha_n(0)/(\alpha_n(0) + \beta_n(0)) \sim 0.32$ in the module representing the n-gate. For the clamp voltage of $V = 5$ mV, $n(t)$ asymptotes to $n_\infty(5) = \alpha_n(5)/(\alpha_n(5)+\beta_n(5))$ which is about 0.4 as shown in Fig. 2.8. As in the case of (2.2), the time constant of the n-gate kinetics is determined by $\tau_n(V_{clamp}) = 1/(\alpha_n(V_{clamp})+\beta_n(V_{clamp}))$ which stays at a constant value when the voltage is clamped at the fixed value V_{clamp}. As one can read from Fig. 2.7-bottom, $\tau_n(5)$ is about 4 ms, and this value is consistent with the rise time of the step response of the n-gate shown in Fig. 2.8-lower. Note that both $n_\infty(V)$ and $\tau_n(V)$ will be no longer constant if the membrane potential V changes dynamically when we consider action potential generations as we will see later in this chapter.

As in the potassium channel with the n-gate kinetics, the sodium channel kinetics can also be described similarly, but using different types of gates. The HH model assumes two types of gates for Na$^+$ channel. They are the m-gate and the h-gate. More specifically, Na$^+$ channel has three m-gates and one h-gate as illustrated in Fig. 2.9.

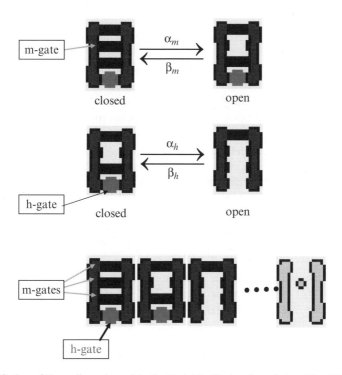

Fig. 2.9 Gating of the sodium channel in the Hodgkin–Huxley formulation. The Na$^+$ channel of the HH model has three m-gates shown as the *dark horizontal bars* from the *top* to the third position of the channel. It has one h-gate which is located at the *bottom* of the channel in this illustration

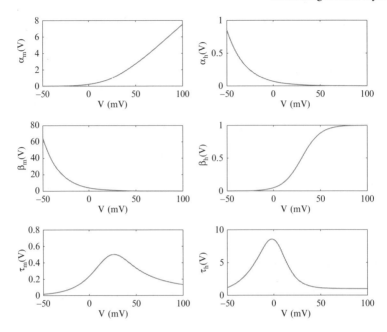

Fig. 2.10 Membrane potential dependency of $\alpha_m(V)$, $\beta_m(V)$ and $\tau_m(V)$ of the m-gate kinetics and $\alpha_h(V)$, $\beta_h(V)$ and $\tau_h(V)$ of the h-gate kinetics for the sodium channel

The m-gate and h-gate have different kinetics. The voltage dependent rate functions determining the transitions between closed and open states of the gates are represented by $\alpha_m(V)$ and $\beta_m(V)$ for each of the m-gates (Fig. 2.9 top row) and by $\alpha_h(V)$ and $\beta_h(V)$ for the h-gate (Fig. 2.9 second row). Note that this diagram at the second row is only for the open–close kinetics of h-gate. That is, open–close of m-gate in this diagram should be neglected. The Na^+ channel is closed at least any one of the four gates is closed as exemplified by three cases in the left part of Fig. 2.9-bottom. It is open if all three m-gates and the h-gate are open, allowing influx and efflux of sodium ions.

As one can observe in Fig. 2.10, $\alpha_m(V)$ and $\beta_m(V)$ are, respectively, monotonically increasing and decreasing functions of V. This is qualitatively the same as the n-gate of the potassium channel. The $\alpha_h(V)$ and $\beta_h(V)$ have opposite characteristics, i.e., the $\alpha_h(V)$ and $\beta_h(V)$ are, respectively, monotonically decreasing and increasing functions of V. When the membrane potential V is low, β_m is greater than α_m, and β_h is smaller than α_h. Thus all m-gates tend to be closed and h-gates tend to be open, implying, in total, that the Na^+ channel tends to be in the closed state for low V. When V is high, α_m is greater than β_m, and α_h is smaller than β_h. Thus the m-gates tend to be open, and the h-gates tend to be closed. Even if all three m-gates are open for a specific single Na^+ channel, the Na^+ channel cannot be in its open state if the h-gate is closed for high values of V. Is there any chance for the Na^+ channel to be open? The answer depends on the dynamics of the m-gate and the h-gate. Here we need to look at the time scales of the m and h gates.

Let us compare the time "constant" functions $\tau_m(V)$ and $\tau_h(V)$ in Fig. 2.10. Although both of them are single humped functions of V, $\tau_h(V)$ is about ten times larger than $\tau_m(V)$ for any value of V. This means that the open–close dynamics of the h-gate are about ten times slower than those of the m-gate. That is, for an increase in the membrane potential from a low potential value for which the h-gates are mostly open, the m-gates open rapidly, leading to the open state of the sodium channel. Then h-gates and thus the sodium channel begins to be closed.

IDE Modeling: *n-m-h* Gating Dynamics with Voltage Clamp: Search ModelDB by "gate" to Find nmh_Gating_Dynamics.isml

$$\frac{dn}{dt} = -\frac{n - n_\infty(V_{clamp})}{\tau_n(V_{clamp})}$$

$$\frac{dm}{dt} = -\frac{m - m_\infty(V_{clamp})}{\tau_m(V_{clamp})}$$

$$\frac{dh}{dt} = -\frac{h - h_\infty(V_{clamp})}{\tau_h(V_{clamp})} \tag{2.9}$$

This IDE model compares the time scales of the n-gate, the m-gate, and the h-gate in response to a step-wise voltage clamp common to these three gate variables by simulating a set of three gating equations in (2.9). This can be modeled on *insilico*IDE by preparing these gating models on the canvas and then providing the output signal of the voltage step generator to each of the gating models as shown in Fig. 2.11 upper panel. The clamp voltage for the first 30 ms is 0 mV and that after 30 ms is $V_{clamp} = 20$ mV. The initial values $n(0)$, $m(0)$, and $h(0)$ are set to $n_\infty(0) \sim 0.32$, $m_\infty(0) \sim 0.05$, and $h_\infty(0) \sim 0.59$, respectively. As shown in Fig. 2.11 lower panel, in response to a step change in the clamp voltage, the m-gate increases (opening of the m-gate) rapidly with its time constant $\tau_m(V_{clamp}) = 1/(\alpha_m(V_{clamp}) + \beta_m(V_{clamp}))$ which is about 0.3 ms, followed by the slow increase in the n-gate (opening of the n-gate) and the slow decrease in the h-gate (closing of the h-gate). Thus, the Na^+ channel takes its open state for a short interval between the onset of the step-change in the clamp voltage until the h-gate is closed with its time constant $\tau_h(V_{clamp}) = 1/(\alpha_h(V_{clamp}) + \beta_h(V_{clamp}))$ which is about 5 ms.

2.3.1 Electro-Chemical Potentials for Cell Membranes

Cells contain relatively high concentration of potassium ions K^+ and low concentration of sodium ions Na^+. This situation is established by the mechanism called the *sodium–potassium pump*. The Na^+/K^+ pump actively moves Na^+ and K^+ ions in opposite directions across the membrane, in which three sodium ions are pumped out, and, at the same time, two potassium ions are pumped in. An enzyme called

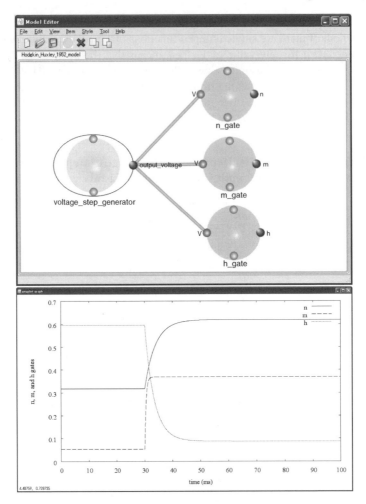

Fig. 2.11 *Upper panel*: The *n-m-h* gate models on *insilico*IDE with the voltage step generator driving the gate models. *Lower panel*: A numerical solution of (2.9) in response to a step-wise voltage clamp. The clamp voltage for the first 30 ms is 0 mV and that after 30 ms is 20 mV. The initial values of $n(0)$, $m(0)$, and $h(0)$ are set to $n_{\infty}(0) \sim 0.32$, $m_{\infty}(0) \sim 0.05$, and $h_{\infty}(0) \sim 0.59$, respectively

Na^+/K^+-ATPase embedded in the membrane is responsible for this pump. Because of this mechanism, the intra-cellular K^+ concentration is kept higher than the extra-cellular K^+ concentration and the intra-cellular Na^+ concentration is kept lower than the extra-cellular Na^+ concentration as illustrated in Fig. 2.12.

When solute ions with different concentrations are separated by the membrane that is permeable to this specific type of ion, the ions move or diffuse across the membrane according to the concentration gradient so that the concentrations are equalized or equilibrated. The statistical mechanics explains this using "a force" whose potential energy is known as *chemical potential*. The diffusion induced by

Fig. 2.12 Intracellular and extracellular spaces

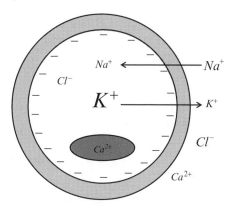

the concentration gradient can also be modeled by Fick's law. Here we consider a one-dimensional space x with a unit of m (meter), which may correspond here to the direction normal to the membrane. In this case, *Fick's law* relates the diffusive flux with a unit of $mol/(m^2 s)$ to the spatial gradient of the concentration as follows;

$$J_c = -D\frac{dC}{dx} \tag{2.10}$$

where J_c represents the flux, which is the amount of the ion that will flow through a unit area within a unit time interval, D the diffusion coefficient with a unit of m^2/s, and C the concentration of the ion with a unit mol/m^3. This just says that the flow is proportional to the intensity of the concentration gradient.

If the ion is electrically charged with its valence z, the electrical force and its potential is also involved. The flow of ions across the membrane is thus driven by concentration gradients and by the electric field. The electric contribution to the flow is determined by *Planck's equation* as follows;

$$J_e = -u\frac{z}{|z|}C\frac{d\phi}{dx} \tag{2.11}$$

where J_e represents the flux induced by the electric field, which is the amount of the charged ion that will flow through a unit area within a unit time interval, u the mobility with a unit of $m^2/(Vs)$, C the concentration of the ion with a unit mol/m^3 as above, and ϕ the electric potential. $z/|z|$ determines the valence-dependent direction of the flow. The mobility u may depend, for example, on the size of the ion. Small (large) ions are easy (hard) to move, leading to a large (a small) mobility. Ions in a viscous liquid may have a small mobility. The Planck's equation just also says that the flow is proportional to the intensity of the electric potential gradient, i.e., the electric field as well as the concentration of the charged ion.

Let us just use, without getting into details, the following relationship (Kubo 1966)

$$D = \frac{uRT}{|z|F} \tag{2.12}$$

which connects the mobility u with the diffusion constant D, where R is the universal gas constant with a unit of $J/(K\,mol)$, and T the absolute temperature in Kelvin, or equivalently

$$u = D\frac{|z|F}{RT}.$$

Then, the total flow of the ion, J, that is transverse to the membrane in the direction of x can be described as;

$$J = J_c + J_e = -D\left(\frac{dC}{dx} + \frac{zF}{RT}C\frac{d\phi}{dx}\right). \tag{2.13}$$

At the electro-chemical equilibrium where the electro-chemical forces are balanced between two sides of the membrane, the net flux J is zero. Thus we obtain

$$-D\left(\frac{dC}{dx} + \frac{zF}{RT}C\frac{d\phi}{dx}\right) = 0, \tag{2.14}$$

so that

$$\frac{1}{C}\frac{dC}{dx} + \frac{zF}{RT}\frac{d\phi}{dx} = 0. \tag{2.15}$$

Let us denote the concentration of the ion inside and outside the membrane, respectively, C_i and C_o, and the electrical potential inside and outside the membrane, respectively, ϕ_i and ϕ_o with assuming that these potentials are constant at either side of the membrane. For any points x_i and x_o inside and outside the membrane, respectively,

$$\int_{x_i}^{x_o} \frac{1}{C(x)}\frac{dC(x)}{dx}dx = -\frac{zF}{RT}\int_{x_i}^{x_o} \frac{d\phi(x)}{dx}dx,$$

which can be rewritten as

$$\int_{C_i}^{C_o} \frac{1}{C(x)}dC(x) = -\frac{zF}{RT}\int_{\phi_i}^{\phi_o} d\phi(x),$$

which yields

$$[\ln C(x)]_{C_i}^{C_o} = \frac{zF}{RT}(\phi_i - \phi_o). \tag{2.16}$$

The electrical potential difference across the membrane is called the *membrane potential*, and we define it as $V = \phi_i - \phi_o$. If we denote the membrane potential V at this equilibrium specifically as E, then (2.16) becomes

$$E = \frac{RT}{zF}\ln\left(\frac{C_o}{C_i}\right). \tag{2.17}$$

Equation (2.17) is called the *Nernst equation*, and E is referred to as the *Nernst potential*.

Much more thorough considerations on this and related issues are available in many literatures such as the one by Keener and Sneyd (2009).

Exercise 2.2. Consider the following intracellular and extracellular ion concentrations. $[K^+]_i = 150\,\text{mM}$, $[K^+]_o = 5\,\text{mM}$, $[Na^+]_i = 15\,\text{mM}$, and $[Na^+]_o = 135\,\text{mM}$. Physical constants are given as $F = 96,500\,\text{C/mol}$, $R = 8.31\,\text{J/mol/K}$, and assume $T = 300\,\text{K}$. Estimate the Nernst potentials for K^+ and Na^+ in the unit of mV.

2.3.2 Electric Circuit Model of Cell Membranes

The flow of ions across the membrane can be viewed as the current. They flow through the channel dominantly, but some part of it is contributed by the capacitant property of the membrane (the lipid bilayer). Because of the capacitant property of the lipid bilayer, it is electrically polarized. The outer and inner surfaces of the membrane are positively and negatively charged, respectively. That is, the membrane as the capacitant stores an amount of electric charge q (Coulomb). With this property, the current–voltage relationship of a cell can be modelled as Fig. 2.13, in which two terminals of the circuit correspond to the inside and the outside of the membrane whose electrical potentials are ϕ_i and ϕ_o, and the difference between them is $V = \phi_i - \phi_o$, the membrane potential. In Fig. 2.13, the outside of the membrane is taken as the ground, i.e., $\phi_o = 0$.

Here, for simplicity, we consider only a single type of the channel through which only a single type of ions flows. The amount of the ionic current may depend on the open fraction of the gates of the channel. The larger the open fraction, the larger is the current. This dependency can be modeled as the resistance R (Ohm) or the

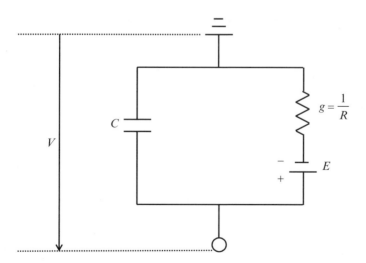

Fig. 2.13 A simple analog electric circuit with resistance and capacitance

conductance $g = 1/R$ (S) of the circuit. A large conductance (or a small resistance) implies the large open fraction of the channel gate. The battery E in Fig. 2.13 corresponds to the Nernst potential for this type of the ion. It drives the channel current if the membrane potential V is deviated from the Nernst potential E.

Using the Ohm's law for the circuit on the right of Fig. 2.13 with its current i_R, the law of Frady for the circuit on the left of Fig. 2.13 with its current i_C, and the Kirchhoff's laws, for a given amount of externally applied current I_{ext}, we have

$$i_R = g(V - E),$$
$$q = CV, \text{ or } i_C \equiv \frac{dq}{dt} = C\frac{dV}{dt},$$
$$I_{ext} = i_R + i_C,$$

yielding the differential equation that governs dynamic change of the voltage difference between the reference and the other side of this RC circuit as follows:

$$C\frac{dV}{dt} = -g(V - E) + I_{ext}, \tag{2.18}$$

where I_{ext} represents the externally injected current. Note that the resistance (conductance) current i_R corresponds to the channel current.

When I_{ext} is constant, using a variable transform as $v = V - E - I_{ext}/g$, (2.18) can be rewritten as

$$\frac{dv}{dt} = -\frac{v}{RC} \tag{2.19}$$

which has exactly the same form as (2.6), and can be solved analytically. Note also that the time constant of the dynamics is RC.

2.4   Hodgkin–Huxley Formulation of Membrane Excitation

As in the electric circuit model described above in (2.18), Hodgkin and Huxley proposed a model of the current through the sodium channel as

$$I_{Na} = \bar{g}_{Na}m^3h(V - E_{Na}) \tag{2.20}$$

where I_{Na} represents the sodium current through all of the Na^+ channels embedded in the membrane, E_{Na} the Nernst potential for the Na^+ ions, and $\bar{g}_{Na}m^3h$ the conductance determined by all Na^+ channels with the gate variables m and h defined previously. Since m and h are the open fractions of the gates when we consider the whole gates in the membrane (and they are the open probability when we consider a specific single channel), they take values between 0 and 1. When all of m and h gates are open, $m = h = 1$, implying the conductance of the sodium channel is $\bar{g}_{Na}$ representing the maximum conductance (the minimum resistance). When all of m

and h gates are closed, $m = h = 0$, implying the conductance of the sodium channel is zero. This means that the resistance of the channel is infinity. In summary, the amount of the sodium current is proportional to the conductance $\bar{g}_{Na}m^3h$ and also to the electric potential $V - E_{Na}$.

Let us consider the form of m^3h for the sodium channel conductance. It reflects the fact that we have modeled the sodium channel with three m-gates and single h-gate. The term m^3 represents the probability that, for a specific channel, all of its three m-gates are open. This is true because we have assumed that the open–close transition of every m-gate is independent of the other m-gates. By the same reason, the m^3 also represents the fraction in the number of the channels for which all three m-gates are open. Together with the h-gate, the term m^3h represents the probability that, for a specific channel, all of its three m-gates and its single h-gate are open. Similarly, the m^3h also represents the fraction in the number of the channels for which all three m-gates and one h-gate are open.

In a similar way, the potassium channel current of the HH model is formulated as

$$I_K = \bar{g}_K n^4 (V - E_K) \tag{2.21}$$

where I_K represents the potassium current through all of the K^+ channels embedded in the membrane, E_K the Nernst potential for the K^+ ions, and $\bar{g}_K n^4$ the conductance for all K^+ channels with the gate variables n defined previously.

The HH model includes three currents. The last one is the leak current I_L which is hypothetized conducting Cl^- ions. The conductance of I_L is assumed as a constant $\bar{g}_L$ which does not depend on the membrane potential, and thus I_L is formulated as

$$I_L = \bar{g}_L (V - E_L) \tag{2.22}$$

where E_L is the Nernst potential of Cl^-.

IDE Modeling: Voltage Clamp Experiment for I_{Na} Current of the HH: Search ModelDB by "Sodium Current" to Find HH_Sodium_Current_Clamp.isml

$$I_{Na} = \bar{g}_{Na}m^3h(V_{clamp} - E_{Na})$$

$$\frac{dm}{dt} = -\frac{m - m_\infty(V_{clamp})}{\tau_m(V_{clamp})}$$

$$\frac{dh}{dt} = -\frac{h - h_\infty(V_{clamp})}{\tau_h(V_{clamp})} \tag{2.23}$$

The electrical potential inside the membrane can be fixed at a constant using a voltage clamp technique. The HH model was derived using such a voltage clamp experiment. Here we simulate (2.23) as in the previous examples of IDE modeling. On the *insilico*IDE, we set two modules representing the voltage step generator and the I_{Na} current of the HH as shown in the upper panel of Fig. 2.14. The initial clamp voltage V_{clamp} is zero, and at $t = 30$ ms, the clamp voltage is re-set to $V_{clamp} = 20$ mV. We simulate a response of the Na$^+$ current to this step change using *insilico*IDE to obtain a transient rapid increase of I_{Na} in the negative direction (inward current) and then relatively slow decrease of the current as shown in Fig. 2.14-lower panel. According to our consideration so far on the gate dynamics, the rapid increase of the current is induced by the rapid opening of the m-gates in response to the onset of the step, and the decrease of the current is by the relatively slow closing of the h-gate. Since the voltage is constant after $t = 30$ ms, the second and the third ODEs in (2.23) can also be solved analytically for $t > 30$ as in (2.4);

$$\bar{m}(t) = m(30) \exp\left(-\frac{t-30}{\tau_m(V_{clamp})}\right) + m_\infty(V_{clamp})\left(1 - \exp\left(-\frac{t-30}{\tau_m(V_{clamp})}\right)\right)$$

$$\bar{h}(t) = h(30) \exp\left(-\frac{t-30}{\tau_h(V_{clamp})}\right) + h_\infty(V_{clamp})\left(1 - \exp\left(-\frac{t-30}{\tau_h(V_{clamp})}\right)\right)$$

Thus we are able to obtain an analytical expression of the step response of I_{Na} as;

$$I_{Na}(t) = \bar{g}_{Na}\bar{m}^3(t)\bar{h}(t)(V_{clamp} - E_{Na}).$$

as the function of time $t > 30$.

The HH model can be summarized as a set of ordinary differential equations as follows;

$$C\frac{dV}{dt} = -\bar{g}_{Na}m^3h(V - E_{Na}) - \bar{g}_K n^4(V - E_K) - \bar{g}_L(V - E_L) + I_{ext}$$

$$\frac{dm}{dt} = \alpha_m(V)(1 - m) - \beta_m(V)m$$

$$\frac{dh}{dt} = \alpha_h(V)(1 - h) - \beta_h(V)h$$

$$\frac{dn}{dt} = \alpha_n(V)(1 - n) - \beta_n(V)n \tag{2.24}$$

By fitting experimental data, Hodgkin and Huxley obtained:

$$
\begin{cases}
\alpha_m(V) = 0.1\dfrac{-V+25}{\exp(\frac{-V+25}{10})-1}, & \beta_m(V) = 4\exp\left(-\dfrac{V}{18}\right) \\[2em]
\alpha_h(V) = 0.07\exp\left(-\dfrac{V}{20}\right), & \beta_h(V) = \dfrac{1}{\exp(\frac{-V+30}{10})+1} \\[2em]
\alpha_n(V) = 0.01\dfrac{-V+10}{\exp(\frac{-V+10}{10})-1}, & \beta_n(V) = 0.125\exp\left(-\dfrac{V}{80}\right) \quad (2.25)
\end{cases}
$$

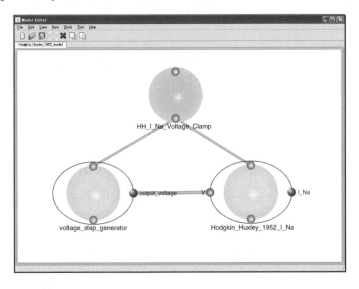

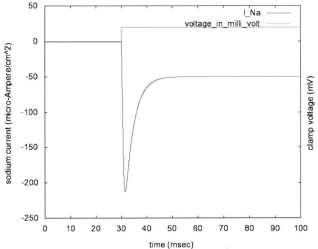

Fig. 2.14 A snapshot of insilicoIDE for a voltage clamp experiment for the HH sodium current, and its simulation outcome

where

I_{ext} externally applied current [μA/cm^2]
V intracellular potential measured from the resting potential,
        positive for the depolarization side [mV]
t time [ms]
m sodium activating variable (dimensionless $0 < m < 1$)
h sodium inactivating variable (dimensionless $0 < h < 1$)
n potassium activating variable (dimensionless $0 < n < 1$)
C membrane capacitance [$= 1\mu$ F/cm^2]

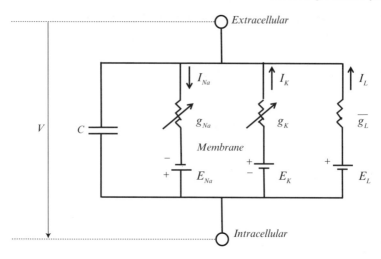

Fig. 2.15 Equivalent electric circuit model of a cellular membrane after Hodgkin and Huxley

E_{Na} sodium equilibrium potential [$= 115\,\text{mV}$]
E_K potassium equilibrium potential [$= -12\,\text{mV}$]
E_L leak equilibrium potential [$= 10.613\,\text{mV}$]
$\bar{g}_{Na}$ maximum sodium conductance [$= 120\,\text{mS}^{-1}/\text{cm}^2$]
$\bar{g}_K$ maximum potassium conductance [$= 36\,\text{mS}^{-1}/\text{cm}^2$]
$\bar{g}_L$ leak conductance [$= 0.3\,\text{mS}^{-1}/\text{cm}^2$]

Figure 2.15 represents the electric circuit model of the HH. Although the circuit includes three channel currents (I_{Na}, I_K, and I_L) all in parallel with the capacitance, it is conceptually the same as the simple circuit of Fig. 2.13.

Let us consider an equilibrium of the HH model before looking at dynamic behaviors of the model. By the equilibrium, we mean a state where a net current, i.e., the sum of all currents, across the membrane is zero so that the membrane potential stays at a constant value referred to as the *resting potential*. Denoting the resting potential as $\bar{V}$, we have

$$\frac{d\bar{V}}{dt} = 0.$$

At the equilibrium, the other dynamic variables, i.e., m, h, and n, do not also change in time, and they stay at, respectively, $\bar{m}$, $\bar{h}$, and $\bar{n}$. Thus,

$$\frac{d\bar{m}}{dt} = 0,$$

$$\frac{d\bar{h}}{dt} = 0,$$

$$\frac{d\bar{n}}{dt} = 0.$$

Therefore, from (2.24), we have

$$-\bar{g}_{Na}\bar{m}^3\bar{h}(\bar{V} - E_{Na}) - \bar{g}_K\bar{n}^4(\bar{V} - E_K) - \bar{g}_L(\bar{V} - E_L) + I_{ext} = 0, \quad (2.26)$$

$$\alpha_m(\bar{V})(1 - \bar{m}) - \beta_m(\bar{V})\bar{m} = 0, \quad (2.27)$$

$$\alpha_h(\bar{V})(1 - \bar{h}) - \beta_h(\bar{V})\bar{h} = 0, \quad (2.28)$$

$$\alpha_n(\bar{V})(1 - \bar{n}) - \beta_n(\bar{V})\bar{n} = 0. \quad (2.29)$$

From (2.27) to (2.29), we have

$$\bar{m} = \frac{\alpha_m(\bar{V})}{\alpha_m(\bar{V}) + \beta_m(\bar{V})} \equiv m_\infty(\bar{V}) \quad (2.30)$$

$$\bar{h} = \frac{\alpha_h(\bar{V})}{\alpha_h(\bar{V}) + \beta_h(\bar{V})} \equiv h_\infty(\bar{V}), \quad (2.31)$$

$$\bar{n} = \frac{\alpha_n(\bar{V})}{\alpha_n(\bar{V}) + \beta_n(\bar{V})} \equiv n_\infty(\bar{V}). \quad (2.32)$$

Moreover, together with (2.26), we have

$$I_{ext} = \bar{g}_{Na}m_\infty(\bar{V})^3 h_\infty(\bar{V})(\bar{V}-E_{Na}) + \bar{g}_K n_\infty(\bar{V})^4(\bar{V}-E_K) + \bar{g}_L(\bar{V}-E_L) \quad (2.33)$$

For a given constant value of I_{ext}, i.e., for a given direct external current injection, one can solve the nonlinear algebraic equation of (2.33) to obtain the unknown resting potential $\bar{V}$. In particular, when $I_{ext} = 0$, we have

$$\bar{V} = \frac{G_{Na}E_{Na} + G_K E_K + G_L E_L}{G_{Na} + G_K + G_L} \quad (2.34)$$

where

$$G_{Na} = \bar{g}_{Na}m_\infty(\bar{V})^3 h_\infty(\bar{V}),$$

$$G_K = \bar{g}_K n_\infty(\bar{V})^4,$$

$$G_L = \bar{g}_L.$$

The resting potential $\bar{V}$ for a given I_{ext} constant can also be obtained graphically by numerically plotting I_{ext} as a function of $\bar{V}$ using (2.33). Reversing the plot to have $\bar{V}$ as a function of I_{ext} will be used to determine the resting potential for a given I_{ext}.

IDE Modeling: The Hodgkin–Huxley Model: Search ModelDB by "Huxley" to Find Hodgkin_Huxley_1952_Model.isml

$$C\frac{dV}{dt} = -\bar{g}_{Na}m^3h(V - E_{Na}) - \bar{g}_K n^4(V - E_K) - \bar{g}_l(V - E_L) + I_{ext}(t)$$

$$\frac{dm}{dt} = -\frac{m - m_\infty(V)}{\tau_m(V)}$$

$$\frac{dh}{dt} = -\frac{h - h_\infty(V)}{\tau_h(V)}$$

$$\frac{dn}{dt} = -\frac{n - n_\infty(V)}{\tau_n(V)} \tag{2.35}$$

This IDE modeling examines the action potential generation in the HH model in response to a single and periodic current pulse stimulations. Equation (2.35) is the same as (2.24), but the right-hand-sides of the ODEs for the gate dynamics are rewritten differently. Figure 2.16-upper shows a snapshot of the IDE modeling of the HH, in which the output of the module representing the periodic current pulse generator (corresponding to the external current I_{ext}) is connected to the module representing the HH model. Here the internal structure of the HH model module is also presented. As described in the later part of this book, an IDE module can represent a tree-like hierarchical structure of modules. This is such an example showing that the HH model is consist of the one module representing the membrane or the membrane potential which is further consist of three modules representing the sodium current, the potassium current, and the leak current. In this case, modules at lower layer of the HH model, up to the ion current level, are expanded and shown. Note that each module representing each of three ion currents has further lower layer modules that represent the gates of the channel. Figure 2.16-middle shows the membrane potential response to a single current pulse injected at time $t = 10$ ms with the accompanied dynamics of the m, h, n gates. The rapid and transient increase of the membrane potential is referred to as the *action potential* as well as the *membrane excitation*. It can be observed that the rapid upstroke (opening) of the m-gate corresponds to the onset of the action potential. The decrease in the membrane potential is caused by the decrease (closing) of the h-gate as well as the increase (opening) of the n-gates, both of which changes slower than the m-gate dynamics as we have analyzed based on the values of the τ_m, τ_h, and τ_n. Note that the opening of the m and h gates induces an inward sodium current $I_{Na} = \bar{g}_{Na}m^3h(V - E_{Na})$ because of the electro-chemical potential (the Nernst potential) of the sodium ions, leading to the increase in the membrane potential. Since $I_K = \bar{g}_K n^4(V - E_K)$ and the Nernst potential of the K^+ ions is opposite to the Na^+ ions, the opening of the n-gates induces an outward current of the potassium ions, leading to the decrease in the membrane potential. Figure 2.16-lower shows a sequence of action potentials in response to the periodic current pulse injections. The full-size action potential is generated in response to the first pulse. The response to

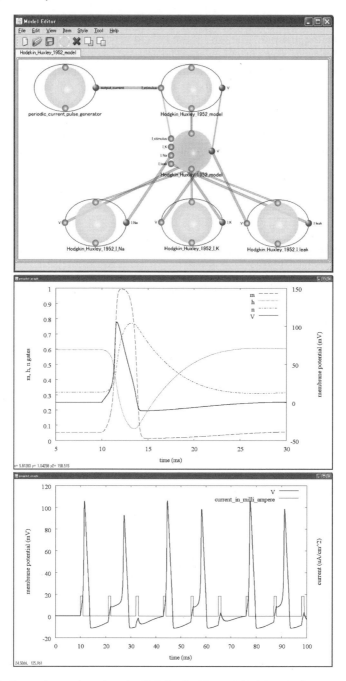

Fig. 2.16 *Upper*: A snapshot of *insilico*IDE for the HH model with pulsatile current injection denoted as $I_{ext}(t)$. *Middle*: An action potential (*solid curve*) in response to a single pulse current with the dynamics of the m-gate (*dashed*), the h-gate (*dotted*), and the n-gate (*dot-dashed*). *Lower*: A response to periodic current pulse stimulations. The period of the pulse sequence for the lower panel is 11 ms. The width and the height of each pulse are 1.0 and 18.445 $\mu A/cm^2$

the second pulse is slightly delayed, and the size of the action potential is slightly smaller than the first one. The third pulse fails to generate the action potential, and so on for the successive current pulses. The delayed response and the failure of the action potential generation are due to the *refractoriness* property of the membrane excitation. That is, once the membrane excites, *excitability* of the membrane decreases for a certain time interval. This is due to the slow recovery dynamics of the h-gate and the n-gate as can be seen from Fig. 2.16-middle. When the h and n gates are recovered, the membrane regains its excitability, allowing full-size action potential responses. The membrane potential at the down-stroke of the action potential goes further below the steady state value that was taken before the action potential generation. This is referred to as *after hyperpolarization* for which the slow closing (inactivation) of the K^+ channel with its slow outward current is responsible.

2.5 FitzHugh–Nagumo Model

Mathematical abstraction of a membrane excitation has been performed by a number of researchers. The most popular one is so-called *FitzHugh–Nagumo (FHN) model* or originally Bonhoeffer–van der Pol (BVP) model (FitzHugh 1961).

$$\frac{dv}{dt} = -v(v-a)(v-b) - w + I_{ext}$$
$$\frac{dw}{dt} = \epsilon(v - cw) \tag{2.36}$$

where $a = 0.139$, $b = 1$, $c = 2.54$, and $\epsilon = 0.008$. The dynamic variables v and w conceptually represent the membrane potential and the refractoriness of the membrane with arbitrary physical units, respectively. I_{ext} represents the externally applied current stimulation.

2.5.1 More About Numerical Integrations of Ordinary Differential Equations

We have looked at the forward and backward Euler methods for numerical integrations, i.e., numerical simulations of ordinary differential equations. Here let us consider them in a more general framework using the FHN model. To this end, (2.36) with $I_{ext} = 0$ is rewritten in a vector form as;

$$\frac{d}{dt} \begin{pmatrix} v \\ w \end{pmatrix} = \begin{pmatrix} -v(v-a)(v-b) - w \\ \epsilon(v - cw) \end{pmatrix} \equiv \begin{pmatrix} f_1(v,w) \\ f_2(v,w) \end{pmatrix} \tag{2.37}$$

More abstraction can be performed by defining $x = (v, w)^T$ and $f(x) = (f_1(x), f_2(x))^T$ to have

$$\frac{dx}{dt} = f(x). \tag{2.38}$$

x is referred to as the *state point* of the model. The state point of the FHN is a two-dimensional vector $x = (v, w)^T$. (As another example, the state point of the HH model is a four-dimensional vector whose components are V, m, h, and n.) For the FHN, the two-dimensional space (plane) of x spanned by v and w is referred as the *phase space* (the *phase plane*) if and only if any point in the space (plane) can uniquely specify a state of the model. As we have seen for the HH model, the equilibrium state of the FHN model at which $dx/dt = 0$ satisfies the following algebraic equation;

$$f(x) = 0. \tag{2.39}$$

A vector-component-wise representation of (2.39) for the FHN is

$$-v(v - a)(v - b) - w = 0 \tag{2.40}$$

$$\epsilon(v - cw) = 0 \tag{2.41}$$

In the $v - w$ phase plane, (2.40) represents the cubic curve $w = -v(v - a)(v - b)$ on which dv/dt is always zero. A set of points consisting this curve is referred to as the *v-nullcline*. Similarly, (2.40) represents the straight line $w = v/c$ on which dw/dt is always zero. A set of points consisting this line is referred to as the *w*-nullcline. See Fig. 2.17 for graphical representation of them. The equilibrium state $\bar{x} = (\bar{v}, \bar{w})^T$ is an intersectional point of these two nullclines. Note that, since *v*-nullcline is nonlinear curve in this model, two nullcline can intersect at multiple points (two or three points at most). In such a case, the model has two or three equilibrium points.

One can see a number of arrows in Fig. 2.17. Each of these arrows is defined for a state point located at the origin of the arrow. More precisely, for any state point x on

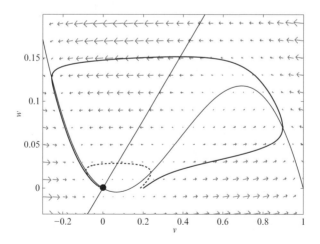

Fig. 2.17 A phase portrait of the FHN model. The cubic shaped *v*-nullcline, straight-line *w* nullcline, the equilibrium point (*closed circle*), and the vector field of the FHN model are shown. Two solution curves started from initial states close to each other are also displayed. One with the *thick curve* moves right-ward corresponding to the onset of the action potential followed by a transient round excursion, eventually settling back to the equilibrium point. In the other with the *dashed curve*, the membrane potential *v* does not increase corresponding to a subthreshold behavior

the phase plane, its arrow is calculated as $f(x)$ which is defined as the left-hand-side of the ODE of (2.38). Such a set of vectors distributed in the phase space is called the *vector field*. As we have studied, the forward Euler scheme with its time step Δt integrates (2.38) to obtain $x(t + \Delta t)$ from a state point $x(t)$ at time t as follows.

$$x(t + \Delta t) = x(t) + \Delta t \cdot f(x(t)). \tag{2.42}$$

This equation means that the state point $x(t)$ moves in the direction of the vector field $f(x(t))$ at $x(t)$ by the amount of $\Delta t \cdot f(x(t))$ to reach at $x(t+\Delta t)$. The smaller the time step Δt, the smaller is the amount of the one step movement, providing better approximation of the real solution of (2.38). In other words, if the time step Δt is not small enough, the numerical solution may be very different from the real solution.

There are many methods for the numerical integration of ordinary differential equations. The fourth-order Runge–Kutta method can provide a better approximation than the forward Euler for a given time step Δt. It is formulated as follows.

$$x(t + \Delta t) = x(t) + \frac{1}{6}\Delta t\,(k_1 + 2k_2 + 2k_3 + k_4) \tag{2.43}$$

where

$$k_1 = f(x(t))$$

$$k_2 = f\left(x(t) + \frac{1}{2}\Delta t k_1\right)$$

$$k_3 = f\left(x(t) + \frac{1}{2}\Delta t k_2\right)$$

$$k_4 = f(x(t) + \Delta t k_3)$$

This method is a fourth-order, meaning that the error per one time step is on the order of $(\Delta t)^5$, and, as it can be shown, the total accumulated error has order of $(\Delta t)^4$. The one step movement in the forward Euler is $\Delta t \cdot f(x(t))$, while it is

$$\Delta t\,(k_1 + 2k_2 + 2k_3 + k_4)/6 \tag{2.44}$$

in the fourth order Runge–Kutta. The one step change in the Euler $\Delta t \cdot f(x(t))$ is determined by a single sample of the vector field at the point $x(t)$, while (2.44) is determined by four samples of the vector field evaluated at the current point $x(t)$ and the other three points $x(t) + \Delta t k_1/2$, $x(t) + \Delta t k_2/2$, and $x(t) + \Delta t k_3$ all of which are close to the point $x(t)$ for a small Δt. This one step movement procedure is summarized in Fig. 2.18. Intuitively, it is obvious that the fourth order

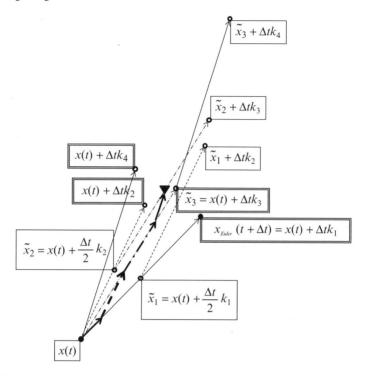

Fig. 2.18 Geometric representation of the fourth-order Runge–Kutta method. The *close circle* is an approximated state point $x(t + \Delta t)$ that is calculated by the forward Euler method from the state point $x(t)$. The *inverted-close triangle* is an approximated state point $x(t + \Delta t)$ obtained by the fourth-order Runge–Kutta method

Runge–Kutta gives better approximation because the one step movement is determined by examining structure of the vector field around the current state point $x(t)$ more in detail by taking a weighted average of those vector fields.

IDE Modeling: Periodically Forced FitzHugh–Nagumo (BVP) Model: Search ModelDB by "BVP" to Find Periodically_Stimulated_BVP_Model.isml

$$\frac{dv}{dt} = c\left(v - \frac{v^3}{3} - w + I_{dc} + I_{ext}(t)\right)$$

$$\frac{dw}{dt} = \frac{1}{c}(v - bw + a) \tag{2.45}$$

where $a = 0.7$, $b = 0.8$, and $c = 3.0$. I_{dc} and I_{ext} are the externally applied current with I_{dc} as the direct current component and $I_{ext}(t)$ as the time-dependent component. This is close to an original version of the FHN model (Nomura et al. 1994; Yoshino et al. 1999) but qualitatively the same as (2.36). Here we examine excitable and oscillatory dynamics of the model for different values of the direct current I_{dc} as well as responses of the model to periodic current stimulations delivered as $I_{ext}(t)$. Figure 2.19 shows action potentials (thin curve) of the BVP model with two different values of I_{dc} in response to a sequence of periodic current pulse stimulations $I_{ext}(t)$ (thick curve) with the period T, the height A, and its width 0.05. The solutions are obtained by integrated with the forward Euler method with the time step 0.0005. In the upper panel, we set $I_{dc} = 0.1$ for which the model behaves as an excitable membrane showing a quiescent (equilibrium or resting) potential when no pulse train is applied for the first 100 ms, and then the model responds in a 3:1 *phase-locked entrainment* to the periodic stimulations ($T = 9.0$ ms and $A = 5.0$) by generating action potentials. In the middle, $I_{dc} = 0.1$ is used as in the upper panel showing a quiescent state for the first 100 ms. Then the model responds irregular-chaotically to the periodic stimulations ($T = 9.0$ ms and $A = 3.1395$). In the lower panel, we set $I_{dc} = 0.347$ for which the model behaves as an oscillatory membrane showing a self-sustained periodic action potentials when no pulse train is applied for the first 100 ms. This solution is a stable *limit cycle*, and it is often consider as a model of repetitive oscillations in *pacemaker cells*. After the onset of the pulse stimulations at 100 ms, the oscillator model responds in a 4:3 phase-locked entrainment to the periodic stimulations ($T = 8.0$ ms and $A = 5.0$). These dynamics can be typically observed in many nonlinear dynamical systems including biological and physiological systems. See Segundo et al. (1991) for a reference of neurophysiological observations associated with the nonlinear dynamics described in this example and thorough analyses of the experimental data. For a model-based analysis related to this example, see Doi and Sato (1995), Nomura et al. (1994), and Yoshino et al. (1999).

2.6   Examples from Cardiac Muscle Cell Models

The heart is one of the most important organs for animals and human. It is composed of cardiac muscle cells. Coordinated contractions of those cells generate heart beats and contractions of the whole heart, allowing blood to circulate in the body. Each contraction of a cardiac cell is controlled and regulated by its cellular excitation. In particular, calcium ions influx through the calcium channel and calcium ions released from calcium stores within a cell induce conformational changes in myofiber proteins, leading to the muscle contraction.

Mechanisms of action potential generation in cardiac cells are basically similar to those in squid axons. Thus, they have been modeled based on the Hodgkin–Huxley formulation and modifications of it. However, cardiac cells, as well as other excitable cells than the squid axons, behave differently if we look at details to some degree. This is natural because they might be responsible for much more complicated physiological roles of life and death than the squid axon. Variety of ion channels and cytoplasmic organella contribute to the action potential generation, muscle contraction, and regulation of them. Understanding the mechanisms is of importance clinically and pharmaceutically because drug-induced sudden cardiac

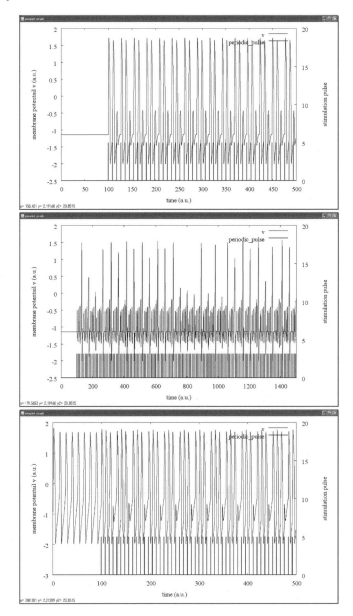

Fig. 2.19 Action potential (*thin curve*) of the BVP model with two different values of I_{dc} in response to a sequence periodic current pulse stimulations $I_{ext}(t)$ (*thick curve*) with the period T, the height A, and the width $w = 0.05$. The application of $I_{ext}(t)$ starts at $t = 100$. Solutions are obtained by the forward Euler method with the time step 0.0005. *Upper*: $I_{dc} = 0.1$ for which the model behaves as an excitable membrane showing a quiescent (resting) potential for the first 100 ms, and then responds in a 3:1 phase-locked entrainment to the periodic stimulations ($T = 9.0$ ms and $A = 5.0$). *Middle*: $I_{dc} = 0.1$ as in the upper for the first 100 ms, and then responds irregular-chaotically to the periodic stimulations ($T = 9.0$ ms and $A = 3.1395$). *Lower*: $I_{dc} = 0.347$ for which the model behaves as an oscillatory membrane (a limit cycle oscillator) showing a self-sustained periodic action potentials for the first 100 ms, and then it responds in a 4:3 phase-locked entrainment to the periodic stimulations ($T = 8.0$ ms and $A = 5.0$)

death and ventricular arrhythmia have arisen as serious public health concerns, in which these arrhythmia could result in withdrawal of non-cardiac drugs. Common drug-induced arrhythmia are due to molecular interaction between chemical compounds of drugs and ion channels (Roden 2004; Sanguinettia and Mitchesonb 2005). Here we briefly look at electrophysiological modeling of cardiac cells through several IDE examples.

A fast inward current of sodium channels is responsible for an onset of cardiac action potential in single cells as in the squid axon. Several kinds of outward potassium currents are also involved, but the time scale of their activations are slower than the axon. Moreover, slow inward currents of calcium channels play important roles, which makes, together with the slow potassium currents, duration of the action potential longer than those of neuronal action potentials. This contributes to stable conduction or propagation of action potentials through a cardiac tissue. The excitation conduction (see next chapter) is a key mechanism of the coordinated muscle contraction starting from a pacemaker site (sino-atrial node) of the heart located at the atrium and then to atrio-ventricular node and the ventricle muscles. The pacemaker cells spontaneously and rhythmically generate action potentials for pacemaking the heart beat.

IDE Modeling: Beeler Reuter Cardiac Ventricular Cell Model: Search ModelDB by "Reuter" to Find Beeler_Reuter_1977_Model.isml

$$\frac{dV_m}{dt} = -\frac{1}{C_m}\left(i_{K1} + i_{x1} + i_{Na} + i_s - i_{external}\right)$$

$$i_{Na} = \left(\bar{g}_{Na}m^3hj + g_{NaC}\right)(V_m - E_{Na})$$

$$i_s = \bar{g}_s df \left(V_m - E_s\right)$$

$$\frac{d[Ca^{2+}]_i}{dt} = -10^{-7}i_s + 0.07\left(10^{-7} - [Ca^{2+}]_i\right)$$

$$E_s = -82.3 - 13.0287\ln[Ca^{2+}]_i \qquad (2.46)$$

where V_m represents the membrane potential, i_{K1} the time-independent potassium outward current, i_{x1} the time-activated outward current, i_{Na} the fast inward sodium current, i_{Ca} the calcium current, $[Ca]_i$ the intracellular calcium concentration, and i_s the slow inward current carried mainly by Ca^{2+}. m, h, j, d, and f are gating variables among other gating variables that are not shown in (2.46). This is a rough description of the Beeler Reuter (BR) model, which is a Hodgkin–Huxley like formulation of a cardiac ventricular cell (Beeler and Reuter 1977). Here we examine its dynamics. This model or several more pioneering models (Noble 1962) triggered a long history up to the present date for understanding cellular excitations and ion channels based on reproduction of electrophysiological observations using mathematical models. Figure 2.20-upper shows a snapshot of the *insilico*IDE canvas

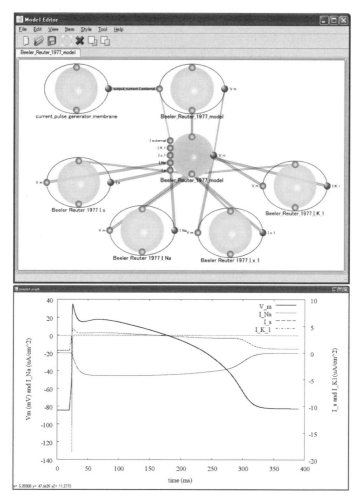

Fig. 2.20 *Upper*: A snapshot of insilicoIDE for the Beeler Reuter (BR) model of a cardiac ventricular cell with a single pulsatile current injection. *Lower*: A response of the model to the single current pulse stimulation. *Thick-solid* represents the membrane potential V_m, *dotted* the sodium current i_{Na}, *dashed* the slow inward current primary carried by Ca^{2+} ions i_s, *dash-dotted* the potassium current i_{K1}

on which a set of modules representing the model is displayed. The action potential waveform is shown in Fig. 2.20-lower with i_{Na}, i_s, and i_{K1}. As in Fig. 2.16-upper, several modules at lower layer of the model, up to the ion current level, are expanded and shown. One can compare that the HH model has three ion current modules in Fig. 2.16 while the BR model has four ion current modules. Despite of the fact that the BR model was constructed only after a decade from the Hodgkin–Huxley with a small modification of the HH, we have already started to feel a hesitation to write down a whole set of equations for the BR model, representing our situation nowadays of how deeply we are struggling against the variety and complexity of the real world modeling. Nevertheless, the whole set of the equations of the BR model is fully described here. Let us emphasize that the description of the BR model here is automatically generated by the *insilico*IDE software which exports the BR model in ISML format into a LaTeX source file to be just pasted in this textbook as follows. The BR model is expressed by 46 equations including the equations for defining parameter values.

I   Beeler_Reuter_1977_Model

This is a capsule module. This model represents the Beeler–Reuter model [Beeler GW, Reuter H (1977) Reconstruction of the action potential of ventricular myocardial fibres. J Physiol 268(1):177–210].

I.1   Beeler_Reuter_1977_I_x_1

This module represents the time-activated outward current model used in the Beeler–Reuter model (Beeler and Reuter 1977). See Sect. V for details.

I.2   Beeler_Reuter_1977_I_Na

This module represents the initial fast inward current model used in the Beeler–Reuter model (Beeler and Reuter 1977). See Sect. III for details.

I.3   Beeler_Reuter_1977_I_K_1

This module represents the time-independent potassium outward current model used in the Beeler–Reuter model (Beeler and Reuter 1977). See Sect. II for details.

I.4   Beeler_Reuter_1977_Model

This model represents the Beeler–Reuter model (Beeler and Reuter 1977).

I.4.1   Physical Quantities

⋆ *shape* : **morphology**
Definition :
   The morphology is assigned to this physical quantity
⋆ V_m : **state**
   The membrane potential taken as inside potential minus outside potential
Definition :

$$\frac{d}{dt}V\_m = \frac{(-((I\_K\_1 + I\_x\_1 + I\_Na + I\_s) - I\_external))}{C\_m} \quad (2.47)$$

with the initial condition

$$V\_m = -84.5737 \quad (2.48)$$

This value is used in

> Module "Beeler_Reuter_1977_model" (Sect. I), physical-quantity ""
> Module "I_s" (Sect. IV.2), physical-quantity "V_m"
> Module "d_gate" (Sect. IV.5), physical-quantity "V_m"
> Module "f_gate" (Sect. IV.1), physical-quantity "V_m"
> Module "I_Na" (Sect. III.1), physical-quantity "V_m"
> Module "m_gate" (Sect. III.3), physical-quantity "V_m"
> Module "h_gate" (Sect. III.2), physical-quantity "V_m"
> Module "j_gate" (Sect. III.4), physical-quantity "V_m"
> Module "I_x_1_bar" (Sect. V.2), physical-quantity "V_m"
> Module "x_1_gate" (Sect. V.1), physical-quantity "V_m"
> Module "I_K_1" (Sect. II.1), physical-quantity "V_m"

⋆ C_m : **static-parameter**
The membrane capacity per unit area
Definition :

$$C_m = 1 \qquad (2.49)$$

⋆ $I_{external}$: **variable-parameter**
This port inputs I_extarnal (current stimulus).
Definition :
The value is given by other module through edge and port
Origin of the value is the Module "current_pulse_generator_membrane" (Sect. VI.1), physical-quantity "$current\_in\_micro\_ampere\_per\_centi\_meter2$"

⋆ I_K_1 : **variable-parameter**
This port inputs I_K_1 (the time-independent potassium outward current).
Definition :
The value is given by other module through edge and port
Origin of the value is the Module "I_K_1" (Sect. II.1), physical-quantity "I_K_1"

⋆ I_x_1 : **variable-parameter**
This port inputs I_x_1 (the time-activated outward current).
Definition :
The value is given by other module through edge and port
Origin of the value is the Module "I_x_1" (Sect. V.3), physical-quantity "I_x_1"

⋆ I_Na : **variable-parameter**
This port inputs I_Na (the initial fast inward current carried primarily by sodium).
Definition :
The value is given by other module through edge and port
Origin of the value is the Module "I_Na" (Sect. III.1), physical-quantity "I_Na"

⋆ I_s : **variable-parameter**
This port inputs I_s (the secondary or slow inward current primarily carried by calcium ions).
Definition :
The value is given by other module through edge and port
Origin of the value is the Module "I_s" (Sect. IV.2), physical-quantity "I_s"

I.5   Beeler_Reuter_1977_I_s

This module represents the secondary or slow inward current model used in the Beeler–Reuter model (Beeler and Reuter 1977). See Sect. IV for details.

II   Beeler_Reuter_1977_I_K_1

This is a capsule module. This module represents the time-independent potassium outward current model used in the Beeler–Reuter model (Beeler and Reuter 1977).

II.1   I_K_1

The time-independent potassium outward current

II.1.1   Physical Quantities

★ *shape* : **morphology**
Definition :
   The morphology is assigned to this physical quantity

★ *I_K_1* : **variable-parameter**
   The time-independent potassium outward current
Definition :

$$I\_K\_1 = 0.35 \left(\frac{4\ (\exp(0.04\ (V\_m + 85)) - 1)}{(\exp(0.08\ (V\_m + 53)) + \exp(0.04\ (V\_m + 53)))} \right.$$

$$\left. + \frac{0.2\ (V\_m + 23)}{(1 - \exp(-0.04\ (V\_m + 23)))} \right) \tag{2.50}$$

This value is used in
   Module "Beeler_Reuter_1977_model" (Sect. I.4), physical-quantity "*I_K_1*"

★ *V_m* : **variable-parameter**
   This port inputs V_m (the membrane potential taken as inside potential minus outside potential).
Definition :
   The value is given by other module through edge and port
   Origin of the value is the Module "Beeler_Reuter_1977_model" (Sect. I.4), physical-quantity "*V_m*"

III   Beeler_Reuter_1977_I_Na

This is a capsule module. This module represents the initial fast inward current model used in the Beeler–Reuter model (Beeler and Reuter 1977).

III.1   I_Na

The initial fast inward current carried primarily by sodium

III.1.1 Physical Quantities

★ *shape* : **morphology**
Definition :
    The morphology is assigned to this physical quantity

★ *g_Na_bar* : **static-parameter**
    The fully activated sodium conductance
Definition :
$$g\_Na\_bar = 4 \tag{2.51}$$

★ *g_NaC* : **static-parameter**
    The background sodium conductance
Definition :
$$g\_NaC = 0.003 \tag{2.52}$$

★ *E_Na* : **static-parameter**
    The reversal potential of Na
Definition :
$$E\_Na = 50 \tag{2.53}$$

★ *I_Na* : **variable-parameter**
    The initial fast inward current carried primarily by sodium
Definition :
$$I\_Na = (g\_Na\_bar \, m \, m \, m \, h \, j + g\_NaC) \,\, (V\_m - E\_Na) \tag{2.54}$$
This value is used in
    Module "Beeler_Reuter_1977_model" (Sect. I.4), physical-quantity "*I_Na*"

★ *V_m* : **variable-parameter**
    This port inputs V_m (the membrane potential taken as inside potential minus outside potential).
Definition :
    The value is given by other module through edge and port
    Origin of the value is the Module "Beeler_Reuter_1977_model" (Sect. I.4), physical-quantity "*V_m*"

★ *m* : **variable-parameter**
    This port inputs m (the proportion of activation gate of I_Na).
Definition :
    The value is given by other module through edge and port
    Origin of the value is the Module "m_gate" (Sect. III.3), physical-quantity "*m*"

★ *h* : **variable-parameter**
    This port inputs h (the proportion of first inactivation gate of I_Na).
Definition :
    The value is given by other module through edge and port
    Origin of the value is the Module "h_gate" (Sect. III.2), physical-quantity "*h*"

★ *j* : **variable-parameter**
    This port inputs j (the proportion of slow inactivation gate of I_Na).
Definition :
    The value is given by other module through edge and port
    Origin of the value is the Module "j_gate" (Sect. III.4), physical-quantity "*j*"

III.2   h_gate

The proportion of first inactivation gate of I_Na

III.2.1   Physical Quantities

$\star$ *shape* : **morphology**
Definition :
   The morphology is assigned to this physical quantity

$\star$ *h* : **state**
   The proportion of first inactivation gate of I_Na
Definition :
$$\frac{d}{dt}h = (alpha\_h \ (1 - h) - beta\_h \ h) \tag{2.55}$$
with the initial condition
$$h = 0.987721 \tag{2.56}$$
This value is used in
   Module "I_Na" (Sect. III.1), physical-quantity "*h*"

$\star$ *alpha_h* : **variable-parameter**
   Opening rate constants of h-gate
Definition :
$$alpha\_h = \frac{(0.126 \ \exp(-0.25 \ (V\_m + 77)) + 0 \ (V\_m + 0))}{(\exp(0 \ (V\_m + 77)) + 0)} \tag{2.57}$$

$\star$ *beta_h* : **variable-parameter**
   Closing rate constants of h-gate
Definition :
$$beta\_h = \frac{(1.7 \ \exp(0 \ (V\_m + 22.5)) + 0 \ (V\_m + 0))}{(\exp(-0.082 \ (V\_m + 22.5)) + 1)} \tag{2.58}$$

$\star$ *V_m* : **variable-parameter**
   This port inputs V_m (the membrane potential taken as inside potential minus outside potential).
Definition :
   The value is given by other module through edge and port
   Origin of the value is the Module "Beeler_Reuter_1977_model" (Sect. I.4), physical-quantity
"*V_m*"

III.3   m_gate

The proportion of activation gate of I_Na

III.3.1   Physical Quantities

$\star$ *shape* : **morphology**
Definition :
   The morphology is assigned to this physical quantity

★ *m* : **state**
The proportion of activation gate of I_Na
Definition :

$$\frac{d}{dt}m = (alpha\_m \ (1 - m) - beta\_m \ m) \tag{2.59}$$

with the initial condition

$$m = 0.010982 \tag{2.60}$$

This value is used in
Module "I_Na" (Sect. III.1), physical-quantity "*m*"

★ *alpha_m* : **variable-parameter**
Opening rate constants of m-gate
Definition :

$$alpha\_m = \frac{(0 \ \exp(0 \ (V\_m + 47)) - (V\_m + 47))}{(\exp(-0.1 \ (V\_m + 47)) - 1)} \tag{2.61}$$

★ *beta_m* : **variable-parameter**
Closing rate constants of m-gate
Definition :

$$beta\_m = \frac{(40 \ \exp(-0.056 \ (V\_m + 72)) + 0 \ (V\_m + 0))}{(\exp(0 \ (V\_m + 72)) + 0)} \tag{2.62}$$

★ *V_m* : **variable-parameter**
This port inputs V_m (the membrane potential taken as inside potential minus outside potential).
Definition :
The value is given by other module through edge and port
Origin of the value is the Module "Beeler_Reuter_1977_model" (Sect. I.4), physical-quantity
"*V_m*"

III.4   *j_gate*

The proportion of slow inactivation gate of I_Na

III.4.1   Physical Quantities

★ *shape* : **morphology**
Definition :
The morphology is assigned to this physical quantity

★ *j* : **state**
The proportion of slow inactivation gate of I_Na
Definition :

$$\frac{d}{dt}j = (alpha\_j \ (1 - j) - beta\_j \ j) \tag{2.63}$$

with the initial condition

$$j = 0.974838 \tag{2.64}$$

This value is used in
Module "I_Na" (Sect. III.1), physical-quantity "*j*"

⋆ *alpha_j* : **variable-parameter**
   Opening rate constants of j-gate
Definition :
$$alpha\_j = \frac{(0.055 \ \exp(-0.25 \ (V\_m + 78)) + 0 \ (V\_m + 0))}{(\exp(-0.2 \ (V\_m + 78)) + 1)} \qquad (2.65)$$

⋆ *beta_j* : **variable-parameter**
   Closing rate constants of j- gate
Definition :
$$beta\_j = \frac{(0.3 \ \exp(0 \ (V\_m + 32)) + 0 \ (V\_m + 0))}{(\exp(-0.1 \ (V\_m + 32)) + 1)} \qquad (2.66)$$

⋆ *V_m* : **variable-parameter**
   This port inputs V_m (the membrane potential taken as inside potential minus outside potential).
Definition :
   The value is given by other module through edge and port
   Origin of the value is the Module "Beeler_Reuter_1977_model" (Sect. I.4), physical-quantity
"*V_m*"

IV   Beeler_Reuter_1977_I_s

This is a capsule module. This module represents the secondary or slow inward current model used
in the Beeler–Reuter model (Beeler and Reuter 1977).

IV.1   f_gate

The proportion of inactivation gate of I_s

IV.1.1   Physical Quantities

⋆ *shape* : **morphology**
Definition :
   The morphology is assigned to this physical quantity

⋆ *f* : **state**
   The proportion of inactivation gate of I_s
Definition :
$$\frac{d}{dt} f = (alpha\_f \ (1 - f) - beta\_f \ f) \qquad (2.67)$$

with the initial condition
$$f = 0.999981 \qquad (2.68)$$

This value is used in
   Module "I_s" (Sect. IV.2), physical-quantity "*f*"

⋆ *alpha_f* : **variable-parameter**
   Opening rate constants of f-gate
Definition :
$$alpha\_f = \frac{(0.012 \ \exp(-0.008 \ (V\_m + 28)) + 0 \ (V\_m + 0))}{(\exp(0.15 \ (V\_m + 28)) + 1)} \qquad (2.69)$$

★ *beta_f* : **variable-parameter**
Closing rate constants of f-gate
Definition :

$$beta\_f = \frac{(0.0065 \ \exp(-0.02 \ (V\_m + 30)) + 0 \ (V\_m + 0))}{(\exp(-0.2 \ (V\_m + 30)) + 1)} \tag{2.70}$$

★ *V_m* : **variable-parameter**
This port inputs V_m (the membrane potential taken as inside potential minus outside potential).
Definition :
The value is given by other module through edge and port
Origin of the value is the Module "Beeler_Reuter_1977_model" (Sect. I.4), physical-quantity
"*V_m*"

IV.2   *I_s*

The secondary or slow inward current primarily carried by calcium ions

IV.2.1   Physical Quantities

★ *shape* : **morphology**
Definition :
The morphology is assigned to this physical quantity

★ *g_s_bar* : **static-parameter**
The fully activated conductance of slow inward ions
Definition :

$$g\_s\_bar = 0.09 \tag{2.71}$$

★ *I_s* : **variable-parameter**
The secondary or slow inward current primarily carried by calcium ions
Definition :

$$I\_s = g\_s\_bar \ d \ f \ (V\_m - E\_s) \tag{2.72}$$

This value is used in
Module "Beeler_Reuter_1977_model" (Sect. I.4), physical-quantity "*I_s*"
Module "Ca_i" (Sect. IV.3), physical-quantity "*I_s*"

★ *V_m* : **variable-parameter**
This port inputs V_m (the membrane potential taken as inside potential minus outside potential).
Definition :
The value is given by other module through edge and port
Origin of the value is the Module "Beeler_Reuter_1977_model" (Sect. I.4), physical-quantity
"*V_m*"

★ *E_s* : **variable-parameter**
This port inputs E_s (the reversal potential of slow inward ions).
Definition :
The value is given by other module through edge and port
Origin of the value is the Module "E_s" (Sect. IV.4), physical-quantity "*E_s*"

★ *d* : **variable-parameter**
This port inputs d (the proportion of activation gate of I_s).

Definition :
The value is given by other module through edge and port
Origin of the value is the Module "d_gate" (Sect. IV.5), physical-quantity "d"

⋆ f : **variable-parameter**
This port inputs f (the proportion of inactivation gate of I_s).
Definition :
The value is given by other module through edge and port
Origin of the value is the Module "f_gate" (Sect. IV.1), physical-quantity "f"

IV.3 Ca_i

The intracellular calcium concentration

IV.3.1 Physical Quantities

⋆ *shape* : **morphology**
Definition :
The morphology is assigned to this physical quantity

⋆ Ca_i : **state**
The intracellular calcium concentration
Definition :
$$\frac{d}{dt}Ca\_i = (-0.0000001\ I\_s + 0.07\ (0.0000001 - Ca\_i))$$ (2.73)
with the initial condition
$$Ca\_i = 0.000000178201$$ (2.74)
This value is used in
Module "E_s" (Sect. IV.4), physical-quantity "Ca_i"

⋆ I_s : **variable-parameter**
This port inputs I_s (the secondary or slow inward current primarily carried by calcium ions).
Definition :
The value is given by other module through edge and port
Origin of the value is the Module "I_s" (Sect. IV.2), physical-quantity "I_s"

IV.4 E_s

The reversal potential of slow inward ions

IV.4.1 Physical Quantities

⋆ *shape* : **morphology**
Definition :
The morphology is assigned to this physical quantity

⋆ E_s : **variable-parameter**
The reversal potential of slow inward ions
Definition :
$$E\_s = (-82.3 - 13.0287\ \ln(Ca\_i))$$ (2.75)

This value is used in
Module "I_s" (Sect. IV.2), physical-quantity "E_s"

⋆ Ca_i : **variable-parameter**
This port inputs Ca_i (the intracellular calcium concentration).
Definition :
The value is given by other module through edge and port
Origin of the value is the Module "Ca_i" (Sect. IV.3), physical-quantity "Ca_i"

IV.5   d_gate

The proportion of activation gate of I_s

IV.5.1   Physical Quantities

⋆ *shape* : **morphology**
Definition :
The morphology is assigned to this physical quantity

⋆ *d* : **state**
The proportion of activation gate of I_s
Definition :

$$\frac{d}{dt}d = (alpha\_d\ (1-d) - beta\_d\ d) \tag{2.76}$$

with the initial condition

$$d = 0.00297072 \tag{2.77}$$

This value is used in
Module "I_s" (Sect. IV.2), physical-quantity "d"

⋆ *alpha_d* : **variable-parameter**
Opening rate constants of d-gate
Definition :

$$alpha\_d = \frac{(0.095\ \exp(-0.01\ (V\_m - 5)) + 0\ (V\_m + 0))}{(\exp(-0.072\ (V\_m - 5)) + 1)} \tag{2.78}$$

⋆ *beta_d* : **variable-parameter**
Closing rate constants of d-gate
Definition :

$$beta\_d = \frac{(0.07\ \exp(-0.017\ (V\_m + 44)) + 0\ (V\_m + 0))}{(\exp(0.05\ (V\_m + 44)) + 1)} \tag{2.79}$$

⋆ *V_m* : **variable-parameter**
This port inputs V_m (the membrane potential taken as inside potential minus outside potential).
Definition :
The value is given by other module through edge and port
Origin of the value is the Module "Beeler_Reuter_1977_model" (Sect. I.4), physical-quantity "V_m"

V   Beeler_Reuter_1977_I_x_1

This is a capsule module. This module represents the time-activated outward current model used in the Beeler–Reuter model (Beeler and Reuter 1977).

V.1   x_1_gate

The proportion of activation gate of I_x_1

V.1.1   Physical Quantities

⋆ *shape* : **morphology**
Definition :
  The morphology is assigned to this physical quantity

⋆ *x_1* : **state**
  The proportion of activation gate of I_x_1
Definition :

$$\frac{d}{dt}x\_1 = (alpha\_x\_1 \ (1 - x\_1) - beta\_x\_1 \ x\_1) \tag{2.80}$$

with the initial condition

$$x\_1 = 0.00562868 \tag{2.81}$$

This value is used in
  Module "I_x_1" (Sect. V.3), physical-quantity "x_1"

⋆ *alpha_x_1* : **variable-parameter**
  Opening rate constants of x_1-gate
Definition :

$$alpha\_x\_1 = \frac{(0.0005 \ \exp(0.083 \ (V\_m + 50)) + 0 \ (V\_m + 0))}{(\exp(0.057 \ (V\_m + 50)) + 1)} \tag{2.82}$$

⋆ *beta_x_1* : **variable-parameter**
  Closing rate constants of x_1-gate
Definition :

$$beta\_x\_1 = \frac{(0.0013 \ \exp(-0.06 \ (V\_m + 20)) + 0 \ (V\_m + 0))}{(\exp(-0.04 \ (V\_m + 20)) + 1)} \tag{2.83}$$

⋆ *V_m* : **variable-parameter**
  This port inputs V_m (the membrane potential taken as inside potential minus outside potential).
Definition :
  The value is given by other module through edge and port
  Origin of the value is the Module "Beeler_Reuter_1977_model" (Sect. I.4), physical-quantity "V_m"

V.2   I_x_1_bar

The maximum time-activated outward current

V.2.1 Physical Quantities

★ *shape* : **morphology**
Definition :
   The morphology is assigned to this physical quantity

★ *I_x_1_bar* : **variable-parameter**
   The maximum time-activated outward current
Definition :

$$I\_x\_1\_bar = \frac{0.8 \ (\exp(0.04 \ (V\_m + 77)) - 1)}{\exp(0.04 \ (V\_m + 35))} \tag{2.84}$$

This value is used in
   Module "I_x_1" (Sect. V.3), physical-quantity "*I_x_1_bar*"

★ *V_m* : **variable-parameter**
   This port inputs V_m (the membrane potential taken as inside potential minus outside potential).
Definition :
   The value is given by other module through edge and port
   Origin of the value is the Module "Beeler_Reuter_1977_model" (Sect. I.4), physical-quantity "*V_m*"

V.3   I_x_1

The time-activated outward current

V.3.1 Physical Quantities

★ *shape* : **morphology**
Definition :
   The morphology is assigned to this physical quantity

★ *I_x_1* : **variable-parameter**
   The time-activated outward current
Definition :

$$I\_x\_1 = I\_x\_1\_bar \ x\_1 \tag{2.85}$$

This value is used in
   Module "Beeler_Reuter_1977_model" (Sect. I.4), physical-quantity "*I_x_1*"

★ *I_x_1_bar* : **variable-parameter**
   This port inputs I_x_1_bar (the maximum time-activated outward current).
Definition :
   The value is given by other module through edge and port
   Origin of the value is the Module "I_x_1_bar" (Sect. V.2), physical-quantity "*I_x_1_bar*"

★ *x_1* : **variable-parameter**
   This port inputs x_1(the proportion of activation gate of I_x_1).
Definition :
   The value is given by other module through edge and port
   Origin of the value is the Module "x_1_gate" (Sect. V.1), physical-quantity "*x_1*"

VI   Current_pulse_generator_membrane

This is a capsule module. current pulse generator module

VI.1   Current_pulse_generator_membrane

Current pulse generator module

VI.1.1   Physical Quantities

⋆ *shape* : **morphology**
Definition :
   The morphology is assigned to this physical quantity

⋆ *pulse_onset_time* : **static-parameter**
   The time instance when the pulse takes place.
Definition :

$$pulse\_onset\_time = 20 \tag{2.86}$$

⋆ *pulse_width* : **static-parameter**
   The pulse width in milli-second.
Definition :

$$pulse\_width = 5 \tag{2.87}$$

⋆ *pulse_height* : **static-parameter**
   The height of the pulse.
Definition :

$$pulse\_height = 10 \tag{2.88}$$

⋆ *initial_ampere* : **static-parameter**
   The current in micro-ampere per centi-meter2 before and after the pulse.
Definition :

$$initial\_ampere = 0 \tag{2.89}$$

⋆ *current_in_micro_ampere_per_centi_meter2* : **variable-parameter**
   The current value changes from an initial value (parameter: initial_ampere) to another value (parameter: pulse_height) at a given onset time (parameter: pulse_onset_time) and after the pulse_width return to the initial value.
Definition :
   When the condition

$$(time \geq pulse\_onset\_time \text{ and } time < (pulse\_onset\_time + pulse\_width)) \tag{2.90}$$

is satisfied, then

$$current\_in\_micro\_ampere\_per\_centi\_meter2 = (pulse\_height$$
$$+ initial\_ampere) \tag{2.91}$$

otherwise

$$current\_in\_micro\_ampere\_per\_centi\_meter2 = initial\_ampere \qquad (2.92)$$

This value is used in
Module "Beeler_Reuter_1977_model" (Sect. I.4), physical-quantity "$I\_external$"

VII Capsule_of_Beeler_Reuter_1977_model

This is a capsule module.

VII.1 Current_pulse_generator_membrane

Current pulse generator module. See Sect. VI for details.

VII.2 Beeler_Reuter_1977_Model

This model represents the Beeler–Reuter model (Beeler and Reuter 1977). See Sect. I for details.

As briefly mentioned above, the calcium ions play dominant roles in the muscle contraction. The cytoplasmic Ca^{2+} concentration in cardiac myocytes is kept extremely low at around $0.1\ \mu M$. An increase of intracellular Ca^{2+} concentration is usually caused by Ca^{2+} influx from the extracellular space and/or Ca^{2+} release from intracellular organelles. Ca^{2+} provides a coupling between electrical excitation and contraction of the cell. The L-type calcium channel on the sarcolemma opens in response to membrane depolarization, allowing Ca^{2+} influx into the dyadic subspace, the narrow space in the cytoplasm between the cell surface and sarcoplasmic reticulum (SR) membranes. When Ca^{2+} influx from the L-type calcium channel diffuses within the dyadic subspace and reaches ryanodine receptors on the SR surface, they trigger Ca^{2+} release from the SR via these receptors, causing further elevation of the cytoplasmic Ca^{2+} concentration. This is known as the *calcium induced calcium release* (CICR) (Fabiato 1985), leading to Ca^{2+} binding to the protein called troponin C for muscle fiber contraction. See Fig. 2.21 for a cartoon of a typical cardiac cell membrane.

IDE Modeling: Faber–Rudy Cardiac Ventricular Cell Model: Search ModelDB by "Rudy" to Find Faber_Rudy_2007_Model.isml

$$C_m \frac{dV_m}{dt} = -I_{FaberRudy} + I_{ext} \qquad (2.93)$$

Faber–Rudy (FR) model (Faber et al. 2007) is one of the most detailed model of a single cardiac ventricular cell. We automatically exported the ISML model of the FR model into a LaTeXsource file as we performed for the BR model using the

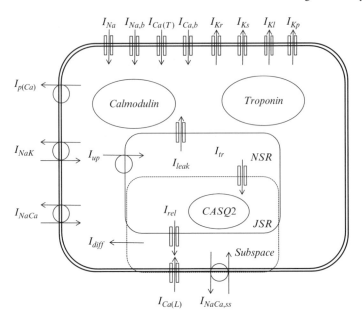

Fig. 2.21 Biophysically detailed model of cardiac ventricular cell membrane

*insilico*IDE, and confirmed that the FR model is described by 333 equations including the ones for defining values of parameters. One could claim that this number (333) is still within a human tractable range for a computational biologist, and it might be true. However, it might also be true that it is not easy task to write every of those equations by hands and to find typos that may be included in the manually produced code. Figure 2.21 illustrates most of the cellular entities included in the FR model. One can see 12 types of ion channels, within which nine are on the plasma membrane and three on the cytoplasma membranes of the intracellular organella, such as the sarcoplasmic reticulum (SR) that stores calcium ions necessary for the muscle contraction. In the FR model, the SR is further divided into two spaces, namely JSR (Junctional SR) and NSR (Network SR). A small space between the cellular membrane and the SR is referred to as the *subspace* or the *dyadic space*. These spaces are illustrated in Fig. 2.21 as square-like boxes located in the cytoplasmic space. Although the model does not explicitly describe the cellular geometry such as the shapes and locations of the SR, the model describes the ion concentration of each of these spaces, such as $[Ca^{2+}]_i$ for the cytoplasmic space and $[Ca^{2+}]_{ss}$ for the dyadic subspace which is determined by

$$\frac{d[Ca^{2+}]_{ss}}{dt} = -\beta_{ss}\left((I_{Ca(L)} - 2I_{NaCa,ss})\frac{A_{cap}}{V_{ss}z_{ca}F} - I_{rel}\frac{V_{JSR}}{V_{ss}} + I_{diff}\right) \quad (2.94)$$

where $I_{Ca(L)}$, $I_{NaCa,ss}$, I_{rel}, and I_{diff} represent, respectively, the currents of the L-type calcium channel, the Na$^+$–Ca^{2+} exchanger in the subspace, the calcium release from the SR into the dyadic subspace inducing the CICR, and the calcium diffusion between the dyadic subspace and the cytoplasmic space. The calcium diffusion is modeled as

$$I_{diff} = \frac{[Ca^{2+}]_{ss} - [Ca^{2+}]_i}{\tau_{diff}} \tag{2.95}$$

as illustrated in Fig. 2.21. This is a simplification of diffusive transportation of Ca^{2+} between these spaces in the ordinary differential equation framework. The model also includes four active ion–ion exchangers (pumps) which are shown as circles on the membranes with one or two arrows representing flows of pumped ions. Figure 2.22-upper shows a snapshot of *insilico*IDE for the FR model with current pulse stimulation generator on the top-left of the panel, while on the right, a set of modules representing the model is displayed where modules at lower layers of the model are expanded still up to the ion current level as in Figs. 2.16 and 2.20. In contrast with Figs. 2.16 and 2.20, the FR model has a lot of lower layer modules because of its complexity in terms of types of ion channels and pumps as well as the intracellular organella. We will look at the FR model in more detail below, in particular, how the L-type calcium channel is modeled. In the lower panels of Fig. 2.22, a response of the FR model to a current pulse stimulation is depicted. The reader is recommended to download the FR model for execution to compare the action potential and the ion currents with those of the BR model in Fig. 2.20.

2.7 Variety of Complex Dynamics of Ion Channels

A number of cellular ion channels have been identified (Hille 2001). They can be classified into several types, or hundred of types if we consider in detail. The classification is made by their gating property and the types of ion that can selectively pass through the channel. Voltage-gated ion channels open or close depending on the voltage across the plasma membrane. Such channel proteins change their structural conformation depending on the membrane potential. In the Hodgkin–Huxley modeling, this voltage-gated property is modeled by the voltage-dependent rate functions $\alpha(V)$ and $\beta(V)$ for each type of the gate. Ligand-gated ion channels open or close depending on binding of ligands to the channel. The term "ligand" is a Latin word meaning "to bind," and in our context, protein molecules that bind to channels or receptors are ligands. Some ligand-gated channels open or close when a ligand outside the cell (extracellular space) binds to the channel, while others when a ligand inside the cell (intracellular space) binds to the channel. Ligand binding causes changes in structural conformation of the channel protein, leading to the opening or closing of the channel. Another type of channels open or close depending on second messengers which are molecules relaying signals from receptors on the cell surface to target molecules inside the cell. This is similar to the action of ligands inside the cell. Ions such as Ca^{2+} could also be considered as second messengers. For example, the L-type calcium channel that we used in the Faber–Rudy model is a

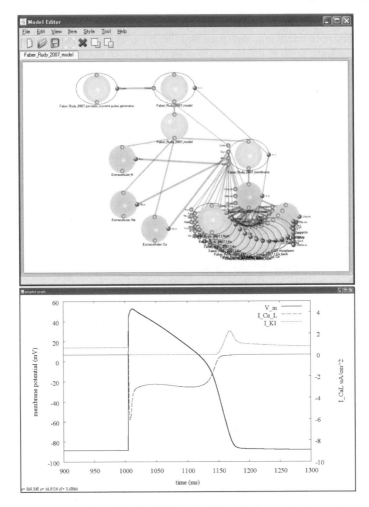

Fig. 2.22 *Upper*: A snapshot of insilicoIDE for the Faber–Rudy model of a cardiac ventricular cell with current pulse stimulations. *Lower*: A response of the model to a current pulse stimulation. *Thick-solid curve* represents the membrane potential V_m, the *dashed* the L-type Ca^{2+} current, and the *dotted* the potassium I_{K1} current. Compare with the BR model in Fig. 2.20

voltage-gated channel, but it is also inactivated by calcium ions. More specifically, Ca^{2+} ions by the influx of L-type calcium channel bind to a type of protein called calmodulin to form a calmodulin–Ca^{2+} complex which binds to the L-type calcium channel and inactivates the channel in a self-feedback manner.

Not all types of channel with such complexity can be modeled by the Hodgkin–Huxley-like gating formulation. This is partially because the HH formulation assumes independence of open–close kinetics of distinguished gates. Moreover, the HH-based gates were introduced hypothetically just to account for channel current dynamics, and they have less correspondence to structures of proteins as real physical entities. A relatively simple example suggesting a limitation of the HH-based gating is the inactivation of the Na$^+$ channel if it is examined in detail. It has been shown that the inactivation of the sodium channel occurs with a larger probability when the channel is open (Armstrong and Bezanilla 1977). This means that the inactivation depends on the states of activation. In terms of the HH gates, this corresponds to a case where closing probability of the h-gate of a single channel depends on if the m-gates of the channel are in their open states or not. This is out of range of the HH gating assumption of stochastic independence among three m-gates and one h-gate, which allows us to multiply three ms and h in obtaining the conductance of the sodium channel. A class of models called *Markov models* can represent more accurately the dependence of a given transition on the occupancy of different states of the channel. See a review by Rudy and Silva (2006) for thorough description on this issue.

Here we only look at several simple examples to compare HH-based gate models and Markov models. Let us consider a simple hypothetical channel with a single open (O) and a single closed (C) state (Fig. 2.23a). The Markov model for this channel is described by the following ODEs.

$$\frac{dC}{dt} = -\alpha C + \beta O, \tag{2.96}$$

$$\frac{dO}{dt} = \alpha C - \beta O, \tag{2.97}$$

where C and O represent the probabilities, respectively, that the state of the channel is in the closed and the open, α and β the transition rates between these two states with their units 1/ms. α and β may be voltage-dependent if the channel is a voltage-gated. In the Markov model, the states C and O represent two different structural conformations of the channel protein. Since the state of the channel is in either closed or open, $C + O = 1$ always holds. For this simple case, we have a HH-based gate model representing exactly the same channel dynamics. Indeed, if we hypothesize a single gate, say m, the probability O can be represented as m, and the probability C as $1 - m$. Then, (2.97) is rewritten as

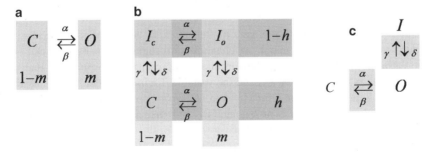

Fig. 2.23 Modeling of ion channel kinetics based on HH-type and Markov-type formulations

$$\frac{d(1-m)}{dt} = -\alpha(1-m) + \beta m,$$

$$\frac{dm}{dt} = \alpha(1-m) - \beta m,$$

in which we have two identical equations. This simple model can be represented by both HH-based gate model and Markov model because there is no state-dependent transition.

Let us consider one more simple example that can also be represented both by HH-based gate model and by Markov model (Fig. 2.23b). We start by a Markov model by assuming four different conformations of the channel. They are closed C, open O, and two inactivated states I_C and I_O. When a channel is in an inactivated state, no ions can flow through the channel, but it is distinguished from the closed state. In this example, I_C is the inactivated state of the closed state, which means that the conformation of I_C is established by a structural change from the closed state. Similarly, I_O is the inactivated state of the open state, which means that the conformation of I_O is established by a structural change from the open state. The channel protein conformations of the inactivated states I_C and I_O can be realized, respectively, only from the closed state C and open state O. That is, these state transitions are state-dependent. However, we assume here that the transition rates between C and I_C and between O and I_O are the same (both γ and δ as in Fig. 2.23b), and those between C and O and between I_C and I_O are the same (both α and β as in Fig. 2.23b). Then the channel dynamics can be represented both by the Markov and the HH-based. The Markov model of this channel is described as

$$\frac{dC}{dt} = \beta O + \delta I_C - (\alpha + \gamma)C, \tag{2.98}$$

$$\frac{dO}{dt} = \alpha C + \delta I_O - (\beta + \gamma)O, \tag{2.99}$$

$$\frac{dI_C}{dt} = \beta I_O + \gamma C - (\alpha + \delta)I_C, \tag{2.100}$$

$$\frac{dI_O}{dt} = \alpha I_C + \gamma O - (\beta + \delta)I_O. \tag{2.101}$$

For the corresponding HH-based gate model, we need to be patient with loosing some detailed information about the channel conformation. We hypothesize two gate variables, m and h, where the variable m represents the probability that the channel state is in either O or I_O, and the variable h represents the probability that the channel state is in either O or C. Then $(1-m)$ represents the probability that the channel state is in either C or I_C, and $(1-h)$ represents the probability that the channel state is in either I_O or I_C. If we are aware of the conformations used in the Markov model, this seems strange, but with the assumption above, this modeling with m and h will work, and the open probability of the channel is represented as mh with the ODEs as follows.

$$\frac{dm}{dt} = \alpha(1 - m) - \beta m,$$

$$\frac{dh}{dt} = \delta(1 - h) - \gamma h.$$

This HH-based formulation is possible because we have assumed that the two horizontal transitions are identical and the two vertical transitions are also identical. Without these particular assumptions, this model essentially involves state dependent transitions, and it cannot be described by the HH formulation.

State transitions for the last example are shown in Fig. 2.23c. This can be modeled only by the Markov scheme, not by the HH-based gating model, because the inactivated state I can be realized only from the open state O. The Markov model of this channel is described as

$$\frac{dC}{dt} = \alpha C - \beta O,$$

$$\frac{dO}{dt} = \alpha C + \delta I - (\beta + \gamma)O,$$

$$\frac{dI}{dt} = \gamma O - \delta I. \tag{2.102}$$

Mathematically speaking, (2.102) for example, can be respresented as

$$\frac{d}{dt}\begin{pmatrix} C \\ O \\ I \end{pmatrix} = \begin{pmatrix} \alpha & -\beta & 0 \\ \alpha & -(\beta + \gamma) & \delta \\ 0 & \gamma & -\delta \end{pmatrix}\begin{pmatrix} C \\ O \\ I \end{pmatrix} \tag{2.103}$$

If the elements of the matrix are constant in a simple case, dynamics of the equation can be analyzed based on the eigenvalues and eigenvectors of the matrix.

IDE Modeling: L-Type Calcium Channel: Search ModelDB by "Rudy" to Find Faber_Rudy_2007_Model.isml

$$I_{Ca(L)} = \bar{I}_{Ca}O, \tag{2.104}$$

$$\bar{I}_{Ca} = P_{Ca}z_{Ca}^2\frac{V_m F^2}{RT} \cdot \frac{\gamma_{Cai}[Ca^{2+}]_{ss}\exp\left(\dfrac{z_{Ca}V_m F}{RT}\right) - \gamma_{Cao}[Ca^{2+}]_o}{\exp\left(\dfrac{z_{Ca}V_m F}{RT}\right) - 1}$$

$$\tag{2.105}$$

where O is the probability that the state of L-Type calcium channel is in open state which is one of 14 states of a Markov-type channel model as described below. This is a model of L-type calcium channel used in the Faber–Rudy model that we examined its action potential above. The reader is asked to draw a transition diagram among those states as an exercise, and compare your diagram with the one shown in the paper by Faber et al. (2007). The Ca^{2+} current through this channel is proportional O as well as $\bar{I}_{Ca}$. The $\bar{I}_{Ca}$ is the ion channel current model based on the famous *Goldman–Hodgkin–Katz (GHK) equation*, which is another way to model membrane channel currents (see below).

$$\frac{dC_0}{dt} = \beta_0 C_1 + \theta C_{0Ca} - (\alpha_0 + \delta)C_0,$$

$$\frac{dC_1}{dt} = \alpha_0 C_0 + \beta_1 C_2 + \theta C_{1Ca} - (\alpha_1 + \beta_0 + \delta)C_1,$$

$$\frac{dC_2}{dt} = \alpha_1 C_1 + \beta_2 C_3 + \theta C_{2Ca} - (\alpha_2 + \beta_1 + \delta)C_2,$$

$$\frac{dC_3}{dt} = \alpha_2 C_2 + \beta_3 O + \omega_f I_{Vf} + \omega_s I_{Vs}$$
$$+ \theta C_{3Ca} - (\alpha_3 + \beta_3 + \gamma_f + \gamma_s + \delta),$$

$$\frac{dO}{dt} = \alpha_3 C_3 + \lambda_f I_{Vf} + \lambda_s I_{Vs} + \theta I_{Ca} - (\beta_3 + \phi_f + \phi_s + \delta)O,$$

$$\frac{dI_{Vf}}{dt} = \gamma_f C_3 + \phi_f O + \omega_{sf} I_{Vs} + \theta I_{VfCa} - (\omega_f + \lambda_f + \omega_{fs} + \delta)I_{Vf},$$

$$\frac{dI_{Vs}}{dt} = \gamma_s C_3 + \phi_s + \omega_{fs} I_{Vf} + \theta I_{VsCa} - (\omega_s + \lambda_s + \omega_{sf} + \delta)I_{Vs},$$

$$\frac{dC_{0Ca}}{dt} = \beta_0 C_{1Ca} + \delta C_0 - (\alpha_0 + \theta)C_{0Ca},$$

$$\frac{dC_{1Ca}}{dt} = \alpha_0 C_{0Ca} + \beta_1 C_{2Ca} + \delta C_1 - (\alpha_1 + \beta_0 + \theta)C_{1Ca},$$

$$\frac{dC_{2Ca}}{dt} = \alpha_1 C_{1Ca} + \beta_2 C_{3Ca} + \delta C_2 - (\alpha_2 + \beta_1 + \theta)C_{2Ca},$$

$$\frac{dC_{3Ca}}{dt} = \alpha_2 C_{2Ca} + \beta_3 I_{Ca} + \omega_f I_{VfCa} + \omega_s I_{VsCa}$$
$$+ \delta C_3 - (\alpha_3 + \beta_2 + \gamma_f + \gamma_s + \theta)C_{3Ca},$$

$$\frac{dC_{Ca}}{dt} = \alpha_3 C_{3Ca} + \lambda_f I_{VfCa} + \lambda_s I_{VsCa} + \delta O - (\beta_3 + \phi_f + \phi_s + \theta)I_{Ca},$$

$$\frac{dI_{VfCa}}{dt} = \gamma_f C_{3Ca} + \phi_f I_{Ca} + \omega_{sf} I_{VsCa} + \delta I_{Vf} - (\omega_f + \lambda_f + \omega_{fs} + \theta) I_{VfCa},$$

$$\frac{dI_{VsCa}}{dt} = \gamma_s C_{3Ca} + \phi_s I_{Ca} + \omega_{fs} I_{VfCa} + \delta I_{Vs} - (\omega_s + \lambda_s + \omega_{sf} + \theta) I_{VsCa}$$

$$(2.106)$$

where most of the Greek letter coefficients are voltage-dependent, and δ is calcium concentration dependent. The reader is asked to simulate the dynamics of this L-type calcium channel by isolating the module representing the L-type calcium channel from the Faber–Rudy model that can be downloaded from the ModelDB. For the IDE operation to perform the module isolation, see the later chapter of this textbook.

Here let us briefly explore the GHK equation to understand how the model of the current $\bar{I}_{Ca}$ in (2.105) is derived. We should be back in the story for the Nernst potential, in which we used the Nernst–Planck equation (2.13). In our context here, we consider a specific part of the cellular membrane (T-tubules) which is faced to the sarcoplasmic reticulum (SR). See Fig. 2.21. The dominant population of L-type calcium channels are typically embedded in the T-tubules of the membrane. $[Ca^{2+}]_{ss}$ used for the model of $\bar{I}_{Ca}$ is the calcium concentration of the dyadic subspace, the small space between the T-tubule membrane and the SR surface, while $[Ca^{2+}]_o$ is the calcium concentration outside the cellular membrane. Then (2.13) in our context becomes

$$J = -D \left(\frac{dc}{dx} + \frac{z_{ca} F}{RT} c \frac{d\phi}{dx} \right) \qquad (2.107)$$

where z_{ca} is the valence of the calcium ion, and J the flux of calcium ions across the membrane. The GHK equation uses so-called the *constant field approximation*, which is a simplification such that the electric field within the membrane (i.e., within the lipid bilayer) can be assumed to be a constant. That is, electric potential is linear function of x as follows.

$$\phi(x) = -Kx$$

where K is a constant, for which the electric field is

$$-\frac{d\phi(x)}{dx} = K \quad \text{(const.)}$$

Denoting electric potential outside ($x = x_o$) and inside ($x = x_i$) the membrane are ϕ_o and ϕ_i, respectively, we have $V_m = \phi_i - \phi_o$. By integrating the electric field from the outside to the inside the membrane as

$$-\int_{x_o}^{x_i} \frac{d\phi(x)}{dx} dx = \int_{x_o}^{x_i} K dx$$

and assuming the thickness of the membrane is $L = x_i - x_o$, we have $-V_m = KL$, and thus

$$K = -\frac{V_m}{L} \qquad (2.108)$$

Using this, (2.107) can be rewritten as the ordinary differential equation with respect to the concentration $c(x)$ as follows.

$$\frac{dc(x)}{dx} = \frac{z_{ca} F V_m}{RTL} c(x) - \frac{J}{D}. \tag{2.109}$$

We integrate (2.109) by spatial coordinate x from x_i to x_o to obtain the flux J as a function of c_i, c_o, and V_m. To this end, (2.109) is rewritten as

$$\frac{dc(x)}{dx} = \frac{z_{ca} F V_m}{RTL} \left(c(x) - \frac{RTLJ}{z_{ca} F V_m D} \right)$$

The variable transform $y(x) = c(x) - RTLJ / (z_{ca} F V_m D)$ is performed to obtain the simpler representation of (2.109) with respect to y as

$$\frac{dy}{y} = \frac{z_{ca} F V_m}{RTL} dx. \tag{2.110}$$

Exercise 2.3. Show that

$$J = \frac{z_{ca} F V_m D}{RTL} \frac{\gamma_{Cai} [Ca^{2+}]_{ss} \exp\left(\dfrac{z_{ca} F V_m}{RT}\right) - \gamma_{Cao} [Ca^{2+}]_o}{\exp\left(\dfrac{z_{ca} F V_m}{RT}\right) - 1} \tag{2.111}$$

by integrating (2.110) from $y_i = c_i - RTLJ / (z_{ca} F V_m D)$ to $y_o = c_o - RTLJ / (z_{ca} F V_m D)$ for the left-hand-side and from x_i to x_o for the right-hand-side, and by using the facts that $x_o - x_i = -L$, and also c_i and c_o can be described by "effective" calcium concentrations inside and outside the membrane, namely, $c_i = \gamma_{Cai} [Ca^{2+}]_{ss}$ and $c_o = \gamma_{Cao} [Ca^{2+}]_o$ for the Faber–Rudy model. γ_{Cai} and γ_{Cao} are referred to as the activity coefficients and represent, respectively, how the calcium ions of the subspace (inside) and outside of the membrane are deviated from the thermodynamic ideal solutes.

The calcium current through the L-type calcium channel $\bar{I}_{Ca}$ when the channel is in the open state ($O = 1$) can then be obtained by multiplying $z_{ca} F$ and (2.111) and denoting D/L as P_{Ca}, leading to (2.105). The ion current modeled by the GHK equation looks relatively complicated and possesses nonlinear dependence on the membrane potential V_m, but both the GHK model and the HH-type linear voltage dependency with the Nernst potential can be derived from the same Planck equation. Thus the use of either the GHK or the Nernst potential is not a matter of degree of modeling refinement. Electrophysiologists choose one of them that can reproduce more satisfactory than the other. Usually, it might be more suitable to use the GHK model rather than the Nernst-potential-based Ohmic model when differences in ion concentrations between inside and outside the membrane are large, in which the current-voltage relationship tends to be nonlinear. Note that the (2.105) is a combination of the GHK current equation and the 14-states Markov-type channel kinetics, which is rather refined and detailed model.

2.8 Variety of Complex Dynamics of Cellular Excitations

Temporal patterns observed in a sequence of action potentials generated by single cells can be complicated. We have seen such a case in the FHN (BVP) with external periodic pulsatile stimulations. In fact, single cells without any external temporally-structured stimulations can autonomously produce complicated patterns of excitation, for example rhythmic bursting excitation.

As such an example, let us consider a model of a pancreatic β cell that shows repetitive, complex, and oscillatory bursting action potentials. Bursting action potentials (spikes) with active (spiking) and silent (non-spiking) phases in the membrane potential of the β cell of the pancreatic islet are induced by glucose, during which insulin is released from the β cell. Insulin is a hormone responsible for the homeostatic mechanism of blood glucose level. It regulates the glucose metabolism by taking up glucose from the blood and by storing metabolites (glycogen) into the liver and muscle cells. In the process of insulin secretion from the β cells, increase of transported glucose into the β cell by a transporter protein (GLUT2) results in increase of ATP released from mitochondrion, which then induces closure of ATP sensitive K^+ channels on the plasma membrane, leading to membrane depolarization. This then triggers opening of voltage-dependent Ca^{2+} channels of the β cell, resulting in increase of intracellular calcium concentration together with Ca^{2+} release from endoplasmic reticulum (ER), which then triggers the secretion of the insulin. See Fig. 2.24. The duration of the active phase and spike frequency within each active phase are altered by the glucose concentration.

For generating bursting dynamics in β cells, as well as in other kinds of bursting cells, fast and slow dynamics interact each other. The fast dynamics are associated with generation of each action potential (spike), typically in a time scale of one or a few milliseconds. The slow dynamics are typically in a time scale of a hundred milliseconds, which are caused by different mechanisms, for example, slow oscillatory changes in cytoplasmic calcium concentration $[Ca^{2+}]_i$, slow dynamics in channel conductance, etc. This situation can be formulated in an abstract form as follows.

$$\frac{dx}{dt} = f(x, y) \tag{2.112}$$

$$\frac{dy}{dt} = \epsilon g(x, y) \tag{2.113}$$

where the system's state is represented by $(n + m)$-dimensional vector $(x, y)^T$ with n-dimensional vector x and m-dimensional vector y. $\epsilon \ll 1$ is a small positive constant. Since ϵ is small, dy/dt is small, meaning that the rate of change in

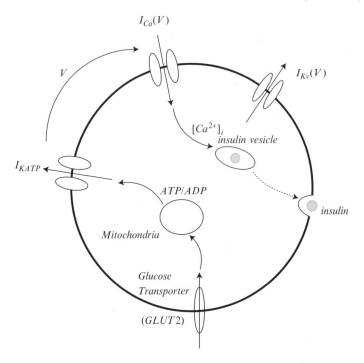

Fig. 2.24 Pancreatic β cell insulin secretion in response to glucose stimulation. Modified from Sherman and Bertram (2005)

the variable y is small and (2.113) represents the slow dynamics component of the whole model of $(x, y)^T$. Equation (2.112) for cellular membrane models exhibiting bursting dynamics describes action potential (spike) generation, that is, the fast dynamics component of the whole model. If we introduce a new time scale $t' = \epsilon t$ (or $t = t'/\epsilon$), the above set of equations can equivalently be rewritten as

$$\epsilon \frac{dx}{dt'} = f(x, y) \tag{2.114}$$

$$\frac{dy}{dt'} = g(x, y) \tag{2.115}$$

In the limit of $\epsilon \to 0$, we have

$$f(x, y) = 0$$

from (2.114), which may be solved formally as

$$x = x(y).$$

That is, x as a function of y. Substituting this into (2.113), we have

$$\frac{dy}{dt} = \epsilon g(x(y), y) \tag{2.116}$$

In many cellular membrane models exhibiting bursting dynamics, (2.116) of the y dynamics shows slow oscillation such as slow calcium oscillations. This slow oscillation may or may not be autonomous, but usually driven by the fast changes in x. For the fast spiking dynamics of x in (2.112), y does not change much. In the limit of $\epsilon \to 0$ in (2.113), y can be regarded as the constant parameter. The spiking dynamics of (2.112) may alter depending on the parameter value of y, either repetitive spiking behavior or non-spiking non-depolarized behavior. In non-limiting situation, y is not really constant, but changes slowly. Thus, for some range of y, the fast dynamics exhibit spiking corresponding to the active phase of the burst, and for the remaining range of y, they show no spikes corresponding to the silent phase of the burst. This is a story of so-called the *fast–slow analysis* of bursting. See Rinzel and Lee (1987), Bertram et al. (1995), Izhikevich (2006) for details. For this material, it is required to study bifurcation phenomena in nonlinear dynamical systems.

IDE Modeling: Bursting Oscillation in a Model Pancreatic β Cell: Search ModelDB by "Beta Cell" to Find Keizer_Smolen_1991.isml

$$C\frac{dV}{dt} = -I_{Ca} - I_{Kv} - I_{KATP} \tag{2.117}$$

where I_{Ca} is the calcium current, I_{Kv} the delayed rectifier potassium current, and I_{KATP} the ATP-sensitive potassium current. This pancreatic β-cell model is proposed by Keizer and Smolen (1991). Some details of three currents are as follows:

$$\frac{dJ}{dt} = -\frac{J - J^\infty(V)}{\tau_J(V)},$$

$$\frac{dn}{dt} = -\frac{n - n^\infty(V)}{\tau_n(V)},$$

$$I_{Ca} = \frac{g_{Ca}[Ca^{2+}]_o V}{[1 - \exp\left(\frac{2FV}{RT}\right)]},$$

$$I_{KATP} = G_{KATP}(V - V_K),$$

$$g_{Ca} = \bar{g}_{Ca}\left(PO_f \cdot X_f + m_s^\infty(V)J(1 - X_f)\right),$$

where J is the slow voltage inactivation of the calcium current, n the gating variable for the delayed rectifier potassium current. I_{Ca} is basically modeled by the Goldman–Hodgkin–Katz (GHK) formulation. The first term of the channel conductance g_{Ca} represents the fast spiking component, which is determined by PO_f dynamics (ODE is not shown), the fraction of open calcium channel of its fast dynamics modeled by a Markov-type state transition. X_f is a static constant representing a fraction of the maximum conductance for the fast component. The second term of g_{Ca} represents the slow dynamics of the calcium channel responsible for the remaining fraction of the calcium conductance $(1 - X_f)$, and it is regulated by the slow inactivation gating dynamics of J. This slow component of the calcium current generates the slow oscillation in the membrane potential on which the fast spiking dynamics is superposed in a nonlinear manner, leading to the bursting dynamics of the model. Figure 2.25-upper shows a snapshot of the IDE with this Keizer and Smole model. In the lower panel, bursting dynamics of the membrane potential with the slow gating dynamics of J are simulated by the *insilico*IDE. The reader is asked to reproduce this with your downloaded model.

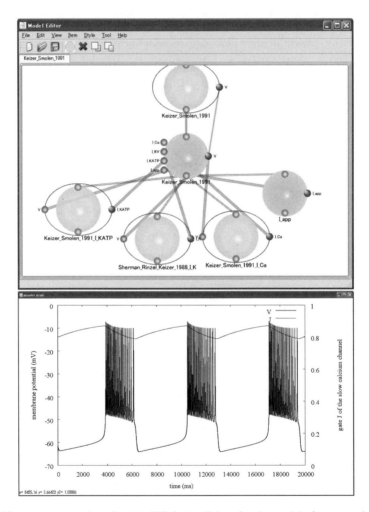

Fig. 2.25 *Upper*: A snapshot of *insilico*IDE for the Keizer–Smolen model of a pancreatic β cell. *Lower*: Bursting action potential of the model. *Thick-solid curve* represents the membrane potential V, the *thin solid* the gate variable J of the slow component of the calcium channel in the right vertical axis scale

2.9 Cellular Calcium Dynamics

Dynamics of intracellular Ca^{2+} are often under tight control by various mechanisms. Ca^{2+} in the dyadic subspace of cardiac cell can be crucial for contraction of myocyte. We have examined the calcium dynamics in the dyadic subspace using the Faber–Rudy (FR) model and its L-type calcium channel modeling, where intracellular Ca^{2+} regulates the amount of Ca^{2+} influx across the sarcolemma through Ca^{2+}-dependent inactivation of the L-type calcium channel. This mechanism is important in the regulation of cardiac *excitation-contraction coupling*. In the FR model, though it is one of the most detailed and complex electrophysiological models of cardiac cellular excitation and calcium dynamics, intracellular calcium dynamics are just modeled by a simple set of ordinary differential equations. See (2.95), for example. If we need to focus the intracellular calcium dynamics more in detail, it will be required to consider spatial dimension and geometry of the intracellular space.

One way to take into account spatial dimension and cytoplasmic geometry is to consider diffusion of calcium ions within the intracellular space. As we learned in (2.10), Fick's law is used here for three-dimensional space with its coordinate $x = (x_1, x_2, x_3)$.

$$J(x,t) = \begin{pmatrix} J_1(x,t) \\ J_2(x,t) \\ J_3(x,t) \end{pmatrix} = -D\nabla[Ca^{2+}](x,t) \equiv -D \begin{pmatrix} \dfrac{\partial[Ca^{2+}](x,t)}{\partial x_1} \\ \dfrac{\partial[Ca^{2+}](x,t)}{\partial x_2} \\ \dfrac{\partial[Ca^{2+}](x,t)}{\partial x_3} \end{pmatrix} \quad (2.118)$$

where $[Ca^{2+}](x,t)$, which is a function of time t and space x, represents the calcium ion concentration at the intracellular space at time t and at the position x. The first, second, and the third component of the J_c represent the flux, the amount of ions that flow through a unit area within a unit time interval in the x_1, x_2, and x_3 direction, respectively. Thus, a change in the amount of calcium ions within in a small cubic of volume $\Delta x_1 \Delta x_2 \Delta x_3$ within a small time interval Δt can be represented as

$$([Ca^{2+}](x, t + \Delta t) - [Ca^{2+}](x,t)) \, \Delta x_1 \Delta x_2 \Delta x_3 =$$
$$[J_1(x_1, x_2, x_3) - J_1(x_1 + \Delta x_1, x_2, x_3)] \, \Delta x_2 \Delta x_3$$
$$[J_2(x_1, x_2, x_3) - J_2(x_1, x_2 + \Delta x_2, x_3)] \, \Delta x_3 \Delta x_1$$
$$[J_3(x_1, x_2, x_3) - J_3(x_1, x_2, x_3 + \Delta x_3)] \, \Delta x_2 \Delta x_3 \quad (2.119)$$

The first term of the right hand side is the amount of ions flowing in the x_1 direction into the volume from the square with its area $\Delta x_2 \Delta x_3$ and flowing out the volume from the square with its area $\Delta x_2 \Delta x_3$. See Fig. 2.26. The second and third terms are defined similarly for the x_2 and x_3 directions, resulting in this equation of balance,

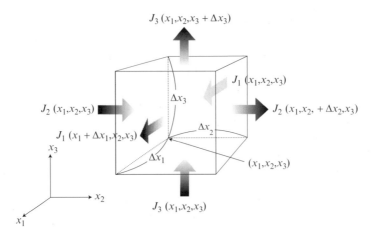

Fig. 2.26 Diffusion of calcium ions in a three dimensional space

referred to as the *equation of continuity*. By taking a limit $\Delta x_i \to 0$ for $i = 1, 2,$ and 3, we have the following equation

$$\frac{\partial [Ca^{2+}](x,t)}{\partial t} = D\nabla^2 [Ca^{2+}](x,t)$$

$$\equiv D \left(\frac{\partial^2 [Ca^{2+}](x,t)}{\partial x_1^2} + \frac{\partial^2 [Ca^{2+}](x,t)}{\partial x_2^2} + \frac{\partial^2 [Ca^{2+}](x,t)}{\partial x_3^2} \right)$$

$$(2.120)$$

referred to as the *diffusion equation*.

Exercise 2.4. Derive (2.120) from (2.119) using (2.118).

If the diffusion of the calcium takes place in a space with no boundary (infinitely large free space), we can assume that $[Ca^{2+}](x,t)$ goes to zero smoothly as $|x| \to \infty$. Thus, (2.120) will be solved with boundary conditions

$$\frac{\partial [Ca^{2+}]}{\partial x_i}(\pm\infty, t) = 0$$

for $i = 1, 2,$ and 3. If the diffusion takes place in a space limited by barriers such as membrane and the surface of cellular organelle, (2.120) should be solved with boundary conditions defined for the surface of the barriers. The calcium ions flowed through the calcium channels into the dyadic subspace of the cardiac myocite diffuse within the dyadic space. In this case, the boundaries are the plasma membrane and the SR surface.

The dyadic space where the calcium ions diffuse are full of proteins such as calcium buffers and calcium binding proteins (Tanskanen et al. 2007). A major calcium binding protein is calmodulin (CaM). It forms a CaM–Ca^{2+} complex when Ca^{2+}

binds to the CaM. CaM and CaM–Ca^{2+} complex also bind to a number of different proteins, by which they regulate various cellular functions. In this way, Ca^{2+} ions diffuse within the dyadic space with interacting with those proteins. Let us consider a diffusion of calcium ions in the presence of calcium buffer molecules B that bind and unbind with Ca^{2+} according to the following chemical reaction.

$$Ca^{2+} + B \underset{k_-}{\overset{k_+}{\rightleftharpoons}} CaB \tag{2.121}$$

Using the *law of mass action*, changes in the calcium concentration caused by this reaction is described as

$$\frac{d[Ca^{2+}]}{dt} = -k_+[Ca^{2+}][B] + k_-[CaB] \tag{2.122}$$

where k_+ and k_- are the reaction rate constants, respectively, for the association (binding) and dissociation (unbinding) of Ca^{2+} with CaM. The first term of the right hand side of (2.122), $[Ca^{2+}][B]$, may be proportional to the number of distinct combinations of the free calciums and the Ca^{2+}-free buffers, and occurrence probability of collisions between free Ca^{2+} and Ca^{2+}-free B may be proportional to this number. If this number of combinations is large, the reaction proceeds rightward frequently to produce CaB complex and the number of free calcium (thus $[Ca^{2+}]$) decreases. Similarly, for the second term, the leftward reaction occurs more frequently as the $[CaB]$ increases, contributing to the increase of $[Ca^{2+}]$. Changes in $[B]$ and $[CaB]$ can similarly be formulated as

$$\frac{d[B]}{dt} = -k_+[Ca^{2+}][B] + k_-[CaB],$$

$$\frac{d[CaB]}{dt} = k_+[Ca^{2+}][B] - k_-[CaB].$$

In the dyadic space, this type of chemical reactions occur in a position x-dependent manner. Thus, with the diffusion dynamics, the spatio-temporal dynamics of the calcium concentration can be described as

$$\frac{\partial[Ca^{2+}]}{\partial t} = D\nabla^2[Ca^{2+}] - k_+[Ca^{2+}][B] + k_-[CaB] \tag{2.123}$$

$$\frac{\partial[B]}{\partial t} = D_B\nabla^2[B] - k_+[Ca^{2+}][B] + k_-[CaB] \tag{2.124}$$

$$\frac{\partial[CaB]}{\partial t} = D_{CaB}\nabla^2[CaB] + k_+[Ca^{2+}][B] - k_-[CaB] \tag{2.125}$$

where D_B and D_{CaB} are the diffusion coefficients of B and CaB, respectively, and they are smaller than D because the buffers are large molecules. For theoretical analyses of this set of reaction diffusion equations, see Ait-Haddou et al. (2010) for example, in which it is shown that the "effective" calcium diffusion can be slow in comparison with that without the buffers if the buffers are present in excess. The effective calcium diffusion coefficient D_{eff} may be formulated as

$$D_{eff} = \frac{D + \kappa D_B}{1 + \kappa} \tag{2.126}$$

where

$$\kappa = \frac{K_d [B]_T}{(K_d + [Ca^{2+}]_\infty)^2}$$

with $K_d = k^+/k_-$, the dissociation constant, $[B]_T = [B] + [CaB]$, the total buffer concentration, which is assumed to be a constant and spatially uniform here. $[Ca]_\infty$ is the free calcium concentration at the equilibrium state. At high buffer concentrations, i.e., at high concentration of $[B]_T$, there left only small amount of free calcium. Thus, if K_d is relatively small, $\kappa \gg 1$, leading to small D_{eff}. This suggests that seemingly simple diffusion of intracellular calcium could be affected by reactions between the calcium and buffer proteins.

2.10 Bridging Between Different Levels of Modeling: A Preliminary Example

Here we consider a simple example that can bridge between two different levels of modeling for intracellular calcium dynamics. In one level, the calcium dynamics within the dyadic subspace of a cardiac myocyte are modeled by an ordinary differential equation referred to as the ODE model or the cellular level model constructed for a whole cell electrophysiology of cardiac cellular excitation. To this end, we use the Faber–Rudy (FR) model again. In the other level, modeling focuses details of the calcium dynamics within the dyadic subspace, in which a model of the calcium dynamics with a geometrical model of the dyadic subspace is constructed using a *Multi-Agent-Simulation* model (MAS) or *Monte Carlo simulation* methodology, referred to as the MAS model or the subcellular level model.

Dynamic changes in the calcium concentration, $[Ca^{2+}]_{ss}$, within the dyadic subspace are of importance because they involve key processes connecting electrophysiology of each cardiac myocyte and mechanical contraction of the cell. For making the regulations of $[Ca^{2+}]_{ss}$ fine and tight, for example, the L-type calcium channel is both voltage and $[Ca^{2+}]_{ss}$ dependent. Particularly, it is inactivated when $[Ca^{2+}]_{ss}$ increases. The L-type calcium channels on the plasma membrane are located in close vicinity to the corresponding SR surface (see Fig. 2.21) on which the ryanodine receptors are arrayed and they trigger the calcium release necessary for muscle contractions in response to increase of $[Ca^{2+}]_{ss}$ by the calcium influx from the L-type calcium channels. Recent studies show that localized and time-dependent calcium concentration within the dyadic subspace, as well as the global-averaged calcium concentration of the subspace, are engaged to the regulation of kinetics of the L-type calcium channel (Tadross et al. 2008).

The ODE (cellular level) model of subspace calcium dynamics in the FR model are described, as in (2.94), by

$$\frac{d[Ca^{2+}]_{ss}}{dt} = -\beta_{ss}\left((I_{Ca(L)} - 2I_{NaCa,ss})\frac{A_{cap}}{V_{ss}z_{ca}F} - I_{rel}\frac{V_{JSR}}{V_{ss}} + I_{diff}\right), \quad (2.127)$$

where $I_{Ca(L)}$, $I_{NaCa,ss}$, I_{rel} modeled by Markov type receptor kinetics (equations are not shown), and I_{diff} represent, respectively, the currents of the L-type calcium channel with the Markov model as in (2.104), the Na^+–Ca^{2+} exchanger in the subspace, the calcium release from the SR into the dyadic subspace inducing a rapid and large increase of $[Ca^{2+}]_{ss}$ referred to as the calcium induced calcium release (CICR), and the calcium diffusion between the dyadic subspace and the cytoplasmic space. I_{diff} is modeled as

$$I_{diff} = \frac{[Ca^{2+}]_{ss} - [Ca^{2+}]_i}{\tau_{diff}} \quad (2.128)$$

where $[Ca^{2+}]_i$ is the cytoplasmic calcium concentration as in (2.95). There are many subspaces in the cytoplasmic space, each of which is a narrow microspace between the plasma membrane and each of many SR surfaces. In the FR model, however, they are lumped together as a single space (a compartment) whose surface area A_{cap} and volume V_{ss} are set to be equal to those of the sum of all subspaces. The calcium diffusion considered in (2.128) is thus the calcium flow between two compartmental spaces, namely, the single compartment as the dyadic subspace and the cytoplasmic space. The Ca^{2+} flow is determined so that it is proportional to the concentration difference between those two spaces. The constant τ_{diff} determines the speed of this diffusion. In the FR model, the constant $\tau_{diff} = 0.1$ ms is utilized.

The solid curve in Fig. 2.27 shows dynamics of the calcium concentration in the dyadic subspace during a voltage clamp condition obtained by the ODE model. In this simulation, the cytoplasmic calcium concentration is fixed at $[Ca^{2+}]_i = 6.72 \times 10^{-5}$ mM. In response to the voltage clamp step applied at $t = 0$, the L-type calcium channel begins to open, and its calcium influx increases rapidly the $[Ca^{2+}]_{ss}$, which induce the calcium release current I_{rel} from the SR, leading to the CICR and the peak calcium concentration $[Ca^{2+}]_{ss}$ about 0.014 mM. The increase of $[Ca^{2+}]_{ss}$ inactivates the L-type calcium channel and the calcium ions in the dyadic subspace diffuse out into the cytoplasmic subspace, and thus the $[Ca^{2+}]_{ss}$ decreases.

The MAS (subcellular level) model concentrates the calcium diffusion within a single dyadic subspace. The single dyadic subspace is modeled as a microspace between the plasma membrane with its area 200×200 nm and the corresponding SR membrane with the same area. The distance between the plasma membrane and the SR surface is set to 50 nm. On the SR membrane, cubic-like-shaped ryanodine receptors are configured. See Fig. 2.28 for overall model geometry. On the plasma

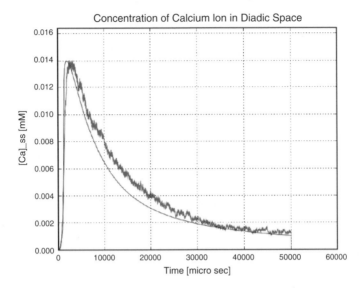

Fig. 2.27 Calcium concentration in the dyadic space obtained by MAS and ODE

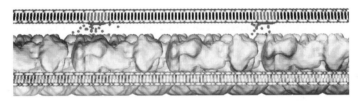

Fig. 2.28 Calcium influx through L-type calcium channels and diffusion within a cellular dyadic space

membrane, four L-type calcium channels are arrayed from which individual Ca^{2+} ion particles are flowed into the subspace if the corresponding L-type channel is in its open state. Each of the L-type calcium channels is modeled using the one used in the FR model, but its kinetics is simulated using a Monte Carlo method implemented as an agent of MAS model. That is, a sample sequence of the state transitions of the Markov model described by (2.106) is realized stochastically using a sequence of pseudo random-numbers without solving the ODEs. For example, let us illustrate this for generating a sample random sequence corresponding to the first ODE of (2.106),

$$\frac{dC_0}{dt} = \beta_0 C_1 + \theta C_{0Ca} - (\alpha_0 + \delta)C_0.$$

This ODE describes changes in the probability that the state of the L-type calcium channel is at C_0, one of the closed states. The state C_0 will move to either C_1 with the rate constant α_0 or to C_{0Ca} with the rate constant δ. The state C_0 is visited

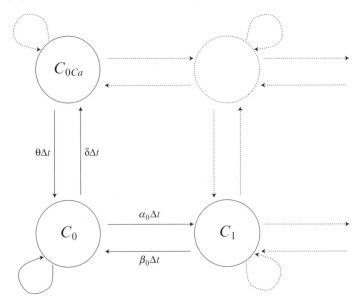

Fig. 2.29 Markov chain representation of a part of the L-type calcium channel

by the transitions from C_1 with the rate constant β_0 and from C_{0Ca} with the rate constant θ. In terms of the MAS or Monte Carlo simulation, this process is modeled by the following Markov chain. See Fig. 2.29. Let us assume that the state is at C_0 at time t, and consider how it moves to other states or it stays at the same state C_0 after the time elapse of Δt. Possible outcomes at $t + \Delta t$ and the corresponding probabilities are as follows.

- Move to the state C_1 with the probability $\alpha_0 \Delta t$
- Move to the state C_{0Ca} with the probability $\delta \Delta t$
- Keep staying at the state C_0 with the probability $1 - \alpha_0 \Delta t - \delta \Delta t$

The state located initially at C_0 may reach the open state O by continuing such stochastics transitions during a single simulation of a given duration. In Fig. 2.27, the time span of a single simulation was set to 50,000 μs. $\Delta t = 1$ ns was used. If the channel state of a L-type calcium channel is in the open state O at a time instant t, a single Ca^{2+} ion particle is created at the position of the pore of the L-type calcium channel in the time interval Δt with the probability 1.92×10^{-2} which corresponds to a single channel current of 6.15×10^{-3} pA. The state transition of each ryanodine receptor is simulated similarly, and if its state is at the open state, a single Ca^{2+} particle is created in Δt at the position of the pore of the ryanodine receptor with the probability 0.8×10^{-2} to simulate the calcium induced calcium release. Some part of transition probabilities in both the L-type calcium channel and the ryanodine receptor are $[Ca^{2+}]_{ss}$-dependent, representing the calcium dependent inactivation for the L-type channels and the calcium induced calcium release from the SR for the ryanodine receptors.

Individual calcium ion particles flowed into the dyadic subspace from the L-type calcium channels and the ryanodine receptors diffuse within the space with the geometry shown in Fig. 2.28. In the MAS-subcellular model, the diffusion of each calcium ion particle is modeled by a simple unbiased Brownian motion. To this end, the dyadic space is discretized into a number of cubics with each side length of Δx. Each calcium ion particle is located at one of their apices at an instant t, and it moves stochastically to one of its six nearest neighbor apices with the identical probabilities, or stay at the same position, for every small time step $\Delta t_D = 0.1$ ns, which is smaller than $\Delta t = 1$ ns used for the state update of the L-type calcium channels and the ryanodine receptors. If a calcium ion particle is going to hit the wall of the dyadic space, i.e., the plasma membrane, the SR surface, or the surface of the ryanodine receptors, by the next time step movement, it is assumed that the ion particle stays at the same position.

For a given time step Δt_D (in this case 0.1 ns), the diffusion speed of each ion particle is determined by the one step length of movement Δx. A large step length corresponds to a large diffusion coefficient D_{Ca} in the diffusion equation. As we used in (2.12), the diffusion coefficient of a calcium ion is determined as

$$D_{Ca} = \frac{u_{Ca} RT}{|z_{Ca}| F}. \tag{2.129}$$

The diffusion coefficient of the calcium within the pure water is then calculated to be 1.58×10^{-12} m^2/ms. However, this value may not be appropriate in our simulation, because there are many calcium binding proteins within the dyadic space which may slow the effective calcium diffusion down as we discussed above. Let us denote the effective diffusion coefficient of the calcium ion particle within the dyadic space $\bar{D}_{Ca}$. Then, as we will see later in (3.18), the diffusion in a single direction from a point calcium source may time-evolve as

$$\exp\left(-\frac{x^2}{2\sigma^2 t}\right),$$

as the function of time of diffusion t, where

$$\sigma = \sqrt{2\bar{D}_{Ca} t}.$$

Therefore, for the small time step Δt_D, the ion particle may move in average about

$$\Delta x \sim \sqrt{2\bar{D}_{Ca} \Delta t_D} \tag{2.130}$$

The noisy curve shown in Fig. 2.27 represents dynamic change in the ensemble averaged $[Ca^{2+}]_{ss}$ within the single dyadic space of the MAS model from 400 sample paths, which show a good coincidence with the solid curve obtained from the ODE cellular level simulation on the FR model. In order to have this coincidence between the subcellular-level MAS model and the cellular-level ODE model, the appropriate one step movement Δx that gives the coincidence was obtained in a trial and error approach. In this example, we got $\Delta x = 2.5$ nm. From (2.130), the corresponding effective diffusion coefficient of the calcium ion is obtained as $\bar{D}_{Ca} = 4.91 \times 10^{-14}$ m^2/ms, which is about 100 times smaller than the calcium diffusion coefficient in the pure water D_{Ca}.

This example gives a rough idea of how different levels (scales) of modeling are inter-related with each other. Since it may not be an easy task to obtain the diffusion coefficient of the calcium within the nanoscale dyadic space, but it could be crucial to understand how the calcium diffusion in the dyadic space is controlled for the excitation-contraction coupling, the approach taken in this example could be beneficial. MAS modeling, though it is computationally time consuming, can provide relatively easy ways to make the modeling more in detail, such as introducing calcium binding proteins in the dyadic space and taking the effect of electric field on the ion movement into account. Simulated data obtained by such mechanism-based detailed modeling should be compared with data obtained by one scale larger ODE modeling as well as by experimental data, leading to deeper understanding of how physiological functions emerge at multiple scales of time and space.

Chapter 3
Dynamics of Cellular Networks

Networks of cells form tissues and organs, where aggregations of cells operate as systems. It is similar to how single cells function as systems of protein networks, where, for example, ion channel currents of a single cell are integrated to produce a whole cell membrane potential. A cell in a network may behave differently from what it does alone. Dynamics of a single cell affect to those of others and vice versa, that is, cells interact with each other. Interactions are made by different mechanisms. Cardiac cells forming a cardiac tissues and heart interact electro-chemically through cell-to-cell connections called *gap junctions*, by which an action potential generated at the sino-atrial node conducts through the heart, allowing coordinated muscle contractions from the atrium to the ventricle. They interact also mechanically because every cell contracts mechanically to produce heart beats. Neuronal cells in the nervous system interact via chemical *synapses*, by which neuronal networks exhibit spatio-temporal spiking dynamics, representing neural information. In a neuronal network in charge of movement control of a musculo-skeletal system, such spatio-temporal dynamics directly correspond to coordinated contractions of a number of skeletal muscles so that a desired motion of limbs can be performed. This chapter illustrates several mathematical techniques through examples from modeling of cellular networks.

3.1 Gap Junctions

A gap junction is an inter-cellular connection between cells. It directly connects the cytoplasm of two adjacent cells, allowing molecules and ions to flow between the cells. One gap junction is composed of two proteins called connexons from opposing cells which come together to form the single inter-cellular gap junction channel. See Fig. 3.1 illustrating coupled three cells via gap junctions.

Let us consider a simple model in which three FHN models as excitable cell membranes connected by gap junctions as in Fig. 3.1. Let us denote the membrane potentials of the first, second, and the third cell models as $v_1(t)$, $v_2(t)$, and $v_3(t)$, respectively. We also denote the corresponding refractoriness variables

T. Nomura and Y. Asai, *Harnessing Biological Complexity*, A First Course in "In Silico Medicine" 1, DOI 10.1007/978-4-431-53880-6_3, © Springer 2011

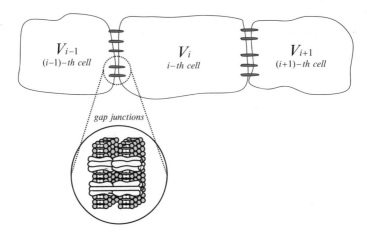

Fig. 3.1 Gap junctions allow electrical interactions between adjacent cells

as $w_1(t)$, $w_2(t)$ and $w_3(t)$, respectively. We consider a case when the external current stimulation I_{ext} is applied only to the first cell model. The most simple way of modeling the gap junction is to assume that it behaves as a passive conductance (a passive resistance) connecting between two terminals (cells). Then, for a given conductance value G of each of the gap junctions, the membrane excitations of the three cell models are described as;

$$\frac{dv_1}{dt} = G(v_2 - v_1) - i_{ion}(v_1, w_1) + I_{ext}(t)$$

$$\frac{dv_2}{dt} = G(v_1 - v_2) + G(v_3 - v_2) - i_{ion}(v_2, w_2)$$

$$\frac{dv_3}{dt} = G(v_2 - v_3) - i_{ion}(v_3, w_3)$$

$$\frac{dw_1}{dt} = \epsilon(v_1 - cw_1)$$

$$\frac{dw_2}{dt} = \epsilon(v_2 - cw_2)$$

$$\frac{dw_3}{dt} = \epsilon(v_3 - cw_3) \tag{3.1}$$

where

$$i_{ion}(v, w) = v(v - a)(v - b) + w$$

IDE Modeling: Three Coupled FitzHugh–Nagumo Model Via Gap Junction: Search ModelDB by "Coupled FHN" to Find Three_Coupled_FHN.isml

$$\frac{dv_1}{dt} = G(v_2 - v_1) - v_1(v_1 - a)(v_1 - b) - w_1 + I_{ext}(t)$$

$$\frac{dv_2}{dt} = G(v_1 - v_2) + G(v_3 - v_2) - v_2(v_2 - a)(v_2 - b) - w_2$$

$$\frac{dv_3}{dt} = G(v_2 - v_3) - v_3(v_3 - a)(v_3 - b) - w_3$$

$$\frac{dw_1}{dt} = \epsilon (v_1 - cw_1)$$

$$\frac{dw_2}{dt} = \epsilon (v_2 - cw_2)$$

$$\frac{dw_3}{dt} = \epsilon (v_3 - cw_3) \tag{3.2}$$

This is an IDE modeling example simulating the excitation conduction in a small network of the FHN cell models. Three FHN model are connected in a series via gap junctions $G = 0.2$, and the cell model at one end is stimulated by a periodic sequence of current pulses. Figure 3.2-upper shows a snapshot of the IDE modeling, in which three modules representing identical FHN models are aligned vertically on the right, and they are connected by lines through the modules representing gap junctions. In Fig. 3.2-lower, the action potential in the one model at one terminal is generated by the periodic current pulse generator located on the top left of the panel, and it causes an excitation of the adjacent cell with a small latency, and then the cell at the other terminal is excited.

3.2 Continuous Approximation and Reaction Diffusion Equations

Let us consider a model of excitation propagation along a spatially one-dimensional excitable medium. Here as a simple example, we again consider the FHN model (2.36) as single cellular components connected by gap junctions. However, here we assume that a system of gap-junction-coupled FHN models as a continuum tissue. Thus we need to consider a spatial dimension x explicitly, and dynamics of the system are described by partial differential equations (PDEs). In this case, the membrane potential and the refractoriness of the tissue at time t and location x are described as $v(x,t)$ and $w(x,t)$ as the functions of t and x. Influences to $v(x,t)$ from neighbors are modeled by the diffusion. Then, we have the following PDEs.

$$\frac{\partial v(x,t)}{\partial t} = D\frac{\partial^2 v(x,t)}{\partial x^2} - i_{ion}\left(v(x,t), w(x,t)\right) + I_{ext}(x,t)$$

$$\frac{\partial w(x,t)}{\partial t} = \epsilon (v(x,t) - cw(x,t)) \tag{3.3}$$

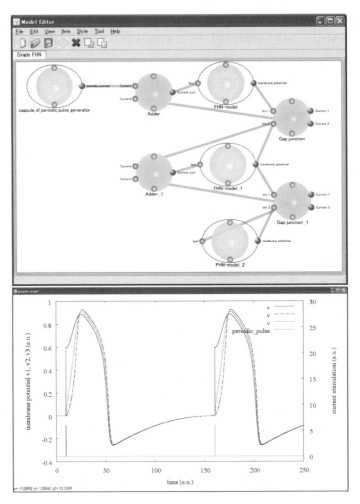

Fig. 3.2 *Upper*: A snapshot of *insilico*IDE for the coupled FHN model (model-1, model-2, and model-3). *Lower*: Conduction of action potential generated in the the model-1 via the gap junctions. *Solid*, *dotted*, and *dashed curves* represent the membrane potentials of v_1, v_2, and v_3, respectively

where

$$i_{ion}(v, w) = v(v - a)(v - b) + w$$
$$= v^3 - (a + b)v^2 + abv + w \qquad (3.4)$$

The first term of the right-hand side of (3.3) for v represents the diffusive current caused by the voltage gradient. The second term is referred to as the reaction term, representing the membrane current flowing through the ion channel. The set up of this *reaction diffusion equation* for the excitation conduction is taken from the literature Rinzel (1977). See also related articles (Glass and Josephson 1995; Nomura and Glass 1996).

3.2.1 Finite Difference Method for Reaction Diffusion Equations

The forward Euler discretization of (3.3) with respect to time t and space x provides a set of ordinary equations that are almost the same as (3.1): Since

$$\frac{\partial^2 v(x,t)}{\partial x^2} \sim \frac{\frac{\partial v(x+\Delta x,t)}{\partial x} - \frac{\partial v(x,t)}{\partial x}}{\Delta x}$$

$$\sim \frac{\frac{v(x+\Delta x,t)-v(x,t)}{\Delta x} - \frac{v(x,t)-v(x-\Delta x,t)}{\Delta x}}{\Delta x}$$

$$= \frac{v(x + \Delta x, t) - 2v(x,t) + v(x - \Delta x, t)}{\Delta x^2}$$

$$= \frac{1}{\Delta x^2}\left(v(x + \Delta x, t) - v(x,t)\right) + \frac{1}{\Delta x^2}\left(v(x - \Delta x, t) - v(x,t)\right),$$

we have equations describing the time evolution of $v(x,t)$ and $w(x,t)$ as;

$$\frac{v(x,t + \Delta t) - v(x,t)}{\Delta t} = \frac{D}{\Delta x^2}\left(v(x + \Delta x, t) - v(x,t)\right)$$

$$+ \frac{D}{\Delta x^2}\left(v(x - \Delta x, t) - v(x,t)\right) - i_{ion}(v(x,t), w(x,t))$$

$$(3.5)$$

$$\frac{w(x,t + \Delta t) - w(x,t)}{\Delta t} = \epsilon\left(v(x,t) - cw(x,t)\right) \qquad (3.6)$$

Equation (3.5) is just the same as the second equation of the Forward Euler discretized version of (3.1) if we identify $v(x - \Delta x, t)$ as v_1, $v(x,t)$ as v_2, $v(x + \Delta x)$ as v_3, and $D/\Delta x^2$ as G.

3.2.2 Cable Equations and Compartment Models

Axonal membranes of neurons are active transmission lines or cables for neuronal signals. The signal carriers are sequences of action potentials. The squid giant axon used for the Hodgkin–Huxley equation is such an axonal membrane. Indeed, the squid giant axon is used to conduct a signal generated at a small brain-like ganglion to the arrays of muscles, leading to coordinated muscle contraction for generating water-jet in escape movement.

As in a cell membrane of somata, an axon of its equilibrium state is electrically polarized, exhibiting a fixed membrane potential which is the difference in the electrical potentials between intra-cellular and extra-cellular spaces. Let us hypothesize a one-dimensional axon cable and a uni-directionally propagating single action potential on the cable. The action potential at a time instant is localized at a small portion of the cable around which the membrane potential is elevated, i.e., the corresponding portion of the membrane is depolarized. Local current circuits

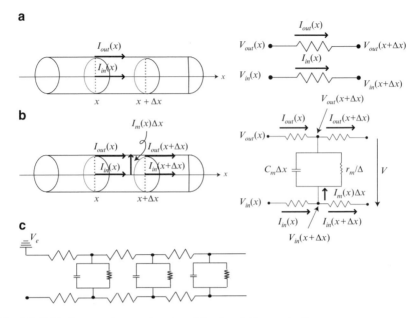

Fig. 3.3 Cable theory and compartment models of excitation conduction

are produced near the action potential. In particular, at the most depolarized portion, the membrane potential starts decreasing, where $dV/dt < 0$ holds. Thus a relatively large amount of capacitance current $C dV/dt$ flows into the membrane from outside. This current then flows according to the membrane potential gradient to the direction of the wave-front of the propagating action potential, and causes an increase in the membrane potential there, which triggers opening of the sodium channel, leading to the excitation of the adjacent portion of the cable membrane. The action potential propagation is produced by chains of this process.

Here we consider an electrically passive cable shown in Fig. 3.3 that is simpler than the active transmission of the action potential. In particular, an one-dimensional passive cable of infinite length is employed to consider several properties of the local membrane current. The active cable with an action potential propagation is considered subsequently.

Let $V_{in}(x, t)$ and V_{out} in volts, respectively, be the electrical potential of the intra-cellular space and the extra-cellular space at time t in second and at a position x in centimeter. Let $I_{in}(x, t)$ and $I_{out}(x, t)$ in ampere, respectively, be the currents that flow longitudinally along the cable inside and outside at the position x, and r_{in} and r_{out} in ohm/cm, respectively, be the resistances for the currents I_{in} and I_{out} per unit length. See Fig. 3.3. Then, by Ohm's law, the voltage drops between the positions x and $x + \Delta x$ due to I_{in} and I_{out} at time t are represented as;

$$V_{in}(x, t) - V_{in}(x + \Delta x, t) = r_{in} \Delta x I_{in}(x, t),$$

$$V_{out}(x, t) - V_{out}(x + \Delta x, t) = r_{out} \Delta x I_{out}(x, t).$$

Thus we have;

$$-\frac{\partial V_{in}(x,t)}{\partial x} = r_{in} I_{in}(x,t), \quad -\frac{\partial^2 V_{in}(x,t)}{\partial x^2} = r_{in}\frac{\partial I_{in}(x,t)}{\partial x}$$

$$-\frac{\partial V_{out}(x,t)}{\partial x} = r_{out} I_{out}(x,t), \quad -\frac{\partial^2 V_{out}(x,t)}{\partial x^2} = r_{out}\frac{\partial I_{out}(x,t)}{\partial x} \tag{3.7}$$

Let $I_m(x,t)$ in ampere/cm be the transmembrane current from the inside to the outside of unit length cable at time t and position x. By the law of local electric charge (current) conservation, we have;

$$I_{in}(x,t) = I_{in}(x + \Delta x, t) + I_m(x,t)\Delta x,$$

$$I_{out}(x,t) = I_{out}(x + \Delta x, t) - I_m(x,t)\Delta x,$$

from which $I_m(x,t)$ can be expressed in two ways as follows;

$$-\frac{\partial I_{in}(x,t)}{\partial x} = I_m(x,t),$$

$$\frac{\partial I_{out}(x,t)}{\partial x} = I_m(x,t) \tag{3.8}$$

Combining (3.7) and (3.8), we have;

$$\frac{\partial^2 V_{in}(x,t)}{\partial x^2} = r_{in} I_m(x,t) \tag{3.9}$$

$$-\frac{\partial^2 V_{out}(x,t)}{\partial x^2} = r_{out} I_m(x,t) \tag{3.10}$$

Defining the membrane potential as $V(x,t) = V_{in}(x,t) - V_{out}(x,t)$, and adding both hands of (3.9) and (3.10), we have;

$$\frac{\partial^2 V(x,t)}{\partial x^2} = (r_{in} + r_{out}) I_m(x,t). \tag{3.11}$$

Let r_m (ohm·cm) be the membrane resistance per unit length of the cable, i.e., the resistance against the current transversing the cable membrane of the unit length. Then the resistance of the cable membrane of the length Δx is $r_m/\Delta x$, meaning that the longer the cable length Δx, the smaller is the resistance. Let c_m (F/cm) be the membrane capacitance of the cable of unit length. Then the capacitance of the cable membrane of the length Δx is $c_m \Delta x$, meaning that the longer the cable length Δx, the wider is the area of its surface and the larger is the capacitance as it is proportional to the length. Thus, the membrane current from the inside to the outside of the cable of the length Δx is the sum of the currents passing through the membrane resistance and the capacitance, and it is represented as;

$$I_m \Delta x = \frac{\Delta x}{r_m} V(x,t) + c_m \Delta x \frac{\partial V(x,t)}{\partial t}$$

or equivalently

$$I_m = \frac{1}{r_m} V(x,t) + c_m \frac{\partial V(x,t)}{\partial t}. \tag{3.12}$$

Finally putting (3.11) and (3.12) together, we have

$$c_m \frac{\partial V(x,t)}{\partial t} - \frac{1}{r_{in} + r_{out}} \frac{\partial^2 V(x,t)}{\partial x^2} + \frac{1}{r_m} V(x,t) = 0 \tag{3.13}$$

$$\tau \frac{\partial V(x,t)}{\partial t} - \lambda^2 \frac{\partial^2 V(x,t)}{\partial x^2} + V(x,t) = 0 \tag{3.14}$$

where

$$\lambda = \sqrt{\frac{r_m}{r_{in} + r_{out}}}$$

or

$$\tau = c_m r_m$$

Equation (3.14) is often referred to as the *cable equation*.

As mentioned above, we assume the cable of infinite length, and $V(x,t) \to 0$ smoothly as $x \to \pm\infty$ to consider the time evolution of a spatial distribution of the voltage along the cable for an impulsive initial distribution of the voltage at time $t = 0$ and at the origin of the position $x = 0$ with the height A volt. This initial condition may be realized by an application of impulsive current injection at $x = 0$ and $t = 0$. We consider a spatio-temporal solution of (3.14) with the initial condition

$$V(x,0) = A\delta(x).$$

where $\delta(x)$ is the Dirac delta function. The Fourier transform technique is utilized for solving a linear partial differential equation, describing a linear reaction diffusion equation by (3.14). Let us define the kernel function of Fourier transform as;

$$\psi_k(x) = \exp(-jkx)$$

where j is the imaginary unit and $k \in \Re$ is the spatial wavenumber (which is equal to $2\pi/\Lambda$ for the wave length Λ). We define the Fourier transform of $V(x,t)$ as

$$\hat{V}(k,t) = \int_{-\infty}^{\infty} V(x,t)\psi_k(x)dx.$$

Using $\hat{V}(k, t)$, $V(x, t)$ can be represented as

$$V(x, t) = \frac{1}{2\pi} \int_{-\infty}^{\infty} \hat{V}(k, t) \bar{\psi}_k(x) dk \tag{3.15}$$

We perform the Fourier transform of both sides of (3.14).

$$\int_{-\infty}^{\infty} \left(\tau \frac{\partial V(x, t)}{\partial t} - \lambda^2 \frac{\partial^2 V(x, t)}{\partial x^2} + V(x, t) \right) \psi_m(x) dx = 0. \tag{3.16}$$

Using (3.15), linearity of (3.14), and partial integral, this integral equality can be further calculated as;

$$\tau \frac{\partial \hat{V}(m, t)}{\partial t} - \lambda^2 \left\{ \left[\frac{\partial V(x, t)}{\partial x} \psi_m(x) \right]_{-\infty}^{\infty} \right.$$

$$\left. + jm \left[V(x, t) \psi_m(x) \right]_{-\infty}^{\infty} - m^2 \hat{V}(m, t) \right\} + \hat{V}(m, t) = 0,$$

thus,

$$\tau \frac{\partial \hat{V}(m, t)}{\partial t} + \left(1 + \lambda^2 m^2 \right) \hat{V}(m, t) = 0.$$

This is the ordinary differential equation for $\hat{V}(m, t)$ with respect to the time t. The solution of the ODE can be obtained as;

$$\hat{V}(m, t) = A \exp\left(-\frac{1 + \lambda^2 m^2}{\tau} t \right) \tag{3.17}$$

since

$$\hat{V}(m, 0) = \int_{-\infty}^{\infty} A\delta(x) \psi_m(x) dx = A.$$

Then, the spatio-temporal evolution of $V(x, t)$ can be calculated as;

$$V(x, t) = \frac{1}{2\pi} \int_{-\infty}^{\infty} \hat{V}(m, t) \bar{\psi}_m(x) dm$$

$$= \frac{A}{2\pi} \int_{-\infty}^{\infty} \exp\left(-\frac{1 + \lambda^2 m^2}{\tau} \right) \exp(jmx) dm$$

$$= \frac{A}{2\pi} \int_{-\infty}^{\infty} \exp\left(-\frac{\lambda^2 t}{\tau} \left[m - j \frac{\tau x}{2\lambda^2 t} \right]^2 - \frac{\tau x^2}{4\lambda^2 t} \right) dm$$

$$= \frac{A}{2\lambda} \sqrt{\frac{\tau}{\pi t}} \exp\left(-\frac{t}{\tau} \right) \exp\left(-\frac{\tau x^2}{4\lambda^2 t} \right) \tag{3.18}$$

This shows that V is distributed with a Gaussian shape whose standard deviation increases as

$$\sqrt{\frac{2t}{\tau}}\lambda. \tag{3.19}$$

Since this solution is for the passive cable, it does not show any conduction of an elevated potential along x axis. Nevertheless, we can gain insight for a distribution of the current I_m across the membrane when the cable has a locally elevated membrane potential at $x = 0$. From (3.11),

$$I_m(x,t) = \frac{1}{r_{in} + r_{out}} \frac{\partial^2 V(x,t)}{\partial x^2}. \tag{3.20}$$

The spatial distribution of V at time t can be represented as

$$V(x,t) = Q(t)\exp\left(-\frac{\tau x^2}{4\lambda^2 t}\right), \tag{3.21}$$

where $Q(t)$ is the time dependent part of (3.18). Using (3.21), (3.20) can be written as

$$I_m(x,t) = \frac{Q(t)\tau}{2\lambda^2 t}\left(\frac{\tau x^2}{2\lambda^2 t} - 1\right)\exp\left(-\frac{\tau x^2}{4\lambda^2 t}\right). \tag{3.22}$$

The locations where $I_m(x,t) = 0$ are found as

$$x = \pm\sqrt{\frac{2t}{\tau}}\lambda, \tag{3.23}$$

which correspond to the inflection points of the Gaussian-shaped voltage distribution. Zeros of the spatial derivative of I_m can be found from the equation

$$\frac{Q(t)\tau x}{2\lambda^4 t^2}\left(1 - \frac{1}{2}\left(\frac{\tau x^2}{2\lambda^2 t} - 1\right)\right) = 0, \tag{3.24}$$

and they are

$$x = 0, \quad \text{and} \quad x = \pm\sqrt{\frac{6t}{\tau}}\lambda. \tag{3.25}$$

The largest inward current across the cable membrane is observed at $x = 0$, most of which is the capacitance current $C dV/dt$. Indeed, since the potential inside the cable is higher than that outside, the current through the resistance around $x = 0$ is outward, but it would be smaller than the capacitance current, leading to the inward current in total. Note that the membrane current $I_m(x,t)$ is inward for positions in

$$\left[-\sqrt{\frac{2t}{\tau}}\lambda, +\sqrt{\frac{2t}{\tau}}\lambda\right].$$

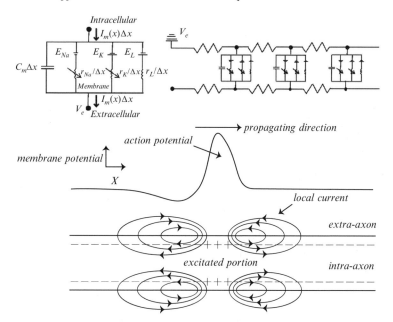

Fig. 3.4 Active cable with its compartment models for excitation conduction

The largest outward currents can be observed at $x = \pm\sqrt{6t/\tau}\lambda$, generating closed loops of the current together with I_{in} and I_{out}.

Exercise 3.1. Draw graphs of $V(x, t)$ and $I_m(x, t)$ as functions of x using a computer for several fixed values of t. The reader may determine λ and τ freely.

Figure 3.4 illustrates an active one-dimensional cable consisting of HH-type active compartments and schematic picture of a conducting action potential along the cable. Spatial distribution of currents, that may correspond to I_m, I_{in}, and I_{out}, forms closed loops.

Exercise 3.2. Try to consider an association between the illustration in Fig. 3.4 for the active cable and the analytical result for the passive cable performed here.

3.2.3 Finite Element Method with Boundary Conditions for Active Transmission Lines

Let us now consider the partial differential equation that describes spatio-temporal dynamics of the membrane potential, i.e., the conduction of the action potential along the one-dimensional active cable as defined in (3.3) again. Let us rewrite (3.3) as;

$$\frac{\partial v}{\partial t} - D\frac{\partial^2 v}{\partial x^2} + I(v, w) = 0$$

$$\frac{\partial w}{\partial t} - \epsilon\left(v(x, t) - cw(x, t)\right) = 0 \tag{3.26}$$

where

$$I(v, w) \equiv i_{ion}(v(x,t), w(x,t))$$

Here we consider the cable of the finite length L. We would like to solve numerically this set of nonlinear differential equations with a given initial condition at $t = 0$ and with a given boundary condition for each at $x = 0$ (the left terminal of the cable) and $x = L$ (the right terminal of the cable). Later in this section, we will look at two cases for the boundary conditions as follows;

Case 1

$$\frac{\partial v}{\partial x}(0, t) = 0, \quad \frac{\partial v}{\partial x}(L, t) = 0.$$

Case 2

$$\frac{\partial v}{\partial x}(0, t) = 0, \quad v(L, t) = \bar{v}.$$

A boundary condition forces a solution of a differential equation to satisfy some constraints at the spatial boundary of the problem. The case 1 here forces a solution $v(x,t)$ to satisfy that its spatial derivative in the direction normal to the boundary is equal to zero at $x = 0$ also at $x = L$. This is called a Neumann boundary condition. In Case 2, the boundary condition at $x = L$ forces a solution $v(x,t)$ to satisfy that its value is equal to a given value (in this case to a constant value $\bar{v}$). This is called a Dirichlet boundary condition.

We approximate the spatial distribution of the membrane potential v and the refractriness w at time t as;

$$\tilde{v}(x,t) = \sum_{k=0}^{n} v_k(t)\phi_k(x)$$

$$\tilde{w}(x,t) = \sum_{k=0}^{n} w_k(t)\phi_k(x) \tag{3.27}$$

where the kernel function $\phi_k(x)$ is defined as;

$$\phi_k(x) = \begin{cases} \dfrac{x - x_{k-1}}{x_k - x_{k-1}} & \text{for } x \in (x_{k-1}, x_k), \\[2mm] \dfrac{x_{k+1} - x}{x_{k+1} - x_k} & \text{for } x \in (x_k, x_{k+1}), \\[2mm] 0 & \text{otherwise} \end{cases}$$

See Fig. 3.5 for a typical shape of the kernel function $\phi_k(x)$, which has non-zero values only at the small localized interval $[x_{k-1}, x_{k+1}]$ and takes zero otherwise (compact support). In particular, $\phi_k(x_k) = 1$. One can notice a similarity, at least conceptually, between (3.27) and (3.15), though the difference in the forms and the support intervals of their kernel functions.

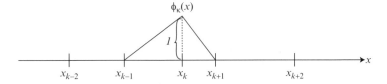

Fig. 3.5 A kernel function $\phi_k(x)$ for the finite element analysis

Then the ion current model of FHN, which includes third order nonlinearity, can be approximated as;

$$i_{ion}(v, w) \sim i_{ion}(\tilde{v}, \tilde{w})$$

$$= \tilde{v}(x,t)\,(\tilde{v}(x,t) - a)\,(\tilde{v}(x,t) - b) + \tilde{w}(x,t)$$

$$= \sum_{i=0}^{n} v_i(t)\phi_i(x) \left\{ \sum_{j=0}^{n} v_j(t)\phi_j(x) - a \right\} \left\{ \sum_{k=0}^{n} v_k(t)\phi_k(x) - b \right\}$$

$$+ \sum_{l=0}^{n} w_l(t)\phi_l(x)$$

$$= \sum_{i=0}^{n}\sum_{j=0}^{n}\sum_{k=0}^{n} v_i(t)v_j(t)v_k(t)\phi_i(x)\phi_j(x)\phi_k(x)$$

$$- a \sum_{i=0}^{n}\sum_{k=0}^{n} v_i(t)v_k(t)\phi_i(x)\phi_k(x)$$

$$- b \sum_{i=0}^{n}\sum_{j=0}^{n} v_i(t)v_j(t)\phi_i(x)\phi_j(x) + ab \sum_{i=0}^{n} v_i(t)\phi_i(x) + \sum_{l=0}^{n} w_l(t)\phi_l(x)$$

$$(3.28)$$

With this, the total approximated current I can be represented as;

$$I(\tilde{v}, \tilde{w}) \equiv i_{ion}(\tilde{v}(x,t), \tilde{w}(x,t))$$

It is desirable for the approximated solutions $\tilde{v}(x,t)$ and $\tilde{w}(x,t)$ in (3.27) to satisfy (3.26) by choosing an appropriate set of coefficients $v_k(t)$ and $w_k(t)$ at time t. To this end, we require the following integral to become zero for every function $\phi_m(x)$ ($m = 1, 2, \ldots, n$):

$$\int_0^L \left(\frac{\partial \tilde{v}}{\partial t} - D\frac{\partial^2 \tilde{v}}{\partial x^2} + I(\tilde{v}, \tilde{w}) \right) \phi_m(x)dx = 0 \qquad (3.29)$$

$$\int_0^L \left(\frac{\partial \tilde{w}}{\partial t} - \epsilon(\tilde{v} - c\tilde{w}) \right) \phi_m(x) dx = 0 \qquad (3.30)$$

Equation (3.29) can also be compared with (3.16).

The first term of (3.29) will be;

$$\int_0^L \left(\frac{\partial \tilde{v}}{\partial t} \right) \phi_m(x) dx$$

$$= \int_0^L \left(\frac{\partial}{\partial t} \sum_{i=0}^n v_i \phi_i(x) \right) \phi_m(x) dx$$

$$= \sum_{i=0}^n \left(\frac{\partial v_i}{\partial t} \right) \int_0^L \phi_i(x) \phi_m(x) dx \qquad (3.31)$$

$$= \frac{x_m - x_{m-1}}{6} \frac{\partial v_{m-1}}{\partial t} + \frac{x_{m+1} - x_{m-1}}{3} \frac{\partial v_m}{\partial t} + \frac{x_{m+1} - x_m}{6} \frac{\partial v_{m+1}}{\partial t} \qquad (3.32)$$

where $v_i(t)$ and $w_i(t)$ are briefly denoted as v_i and w_i, using the following integra for $m \neq 0$ and $m \neq n$;

$$\int_0^L \phi_{m-1}(x) \phi_m(x) dx = \frac{x_m - x_{m-1}}{6}, \quad \text{for } i = m-1 \text{ or } l = m-1$$

$$\int_0^L \phi_m(x) \phi_m(x) dx = \frac{x_{m+1} - x_{m-1}}{3}, \quad \text{for } i = m \text{ or } l = m$$

$$\int_0^L \phi_m(x) \phi_{m+1}(x) dx = \frac{x_{m+1} - x_m}{6}, \quad i = m+1 \text{ or } l = m+1$$

$$\int_0^L \phi_i(x) \phi_m(x) dx = 0, \qquad \text{otherwise} \qquad (3.33)$$

By similar calculations for the integra for $m = 0$ and $m = n$, (3.31) for $m = 0$ becomes

$$\frac{x_1 - x_0}{3} \frac{\partial v_0}{\partial t} + \frac{x_1 - x_0}{6} \frac{\partial v_1}{\partial t} \qquad (3.34)$$

and, for $m = n$,

$$\frac{x_n - x_{n-1}}{6} \frac{\partial v_{n-1}}{\partial t} + \frac{x_n - x_{n-1}}{3} \frac{\partial v_n}{\partial t}. \qquad (3.35)$$

Using a partial integral, the second term of (3.29) is further calculated as;

$$-\int_0^L D\frac{\partial^2 \tilde{v}}{\partial x^2}\phi_m(x)dx$$

$$= -\left[D\frac{\partial \tilde{v}}{\partial x}\phi_m(x)\right]_0^L + \int_0^L D\frac{\partial \tilde{v}}{\partial x}\frac{d\phi_m(x)}{dx}dx$$

$$= -D\frac{\partial \tilde{v}}{\partial x}(L,t)\phi_m(L) + D\frac{\partial \tilde{v}}{\partial x}(0,t)\phi_m(0) + \sum_{i=0}^{n} Dv_i \int_0^L \frac{d\phi_i(x)}{dx}\frac{d\phi_m(x)}{dx}dx$$

$$(3.36)$$

The subscript i in the last term of (3.36) runs from 0 to n, and for different combinations of i and m,

$$\int_0^L \frac{d\phi_{m-1}(x)}{dx}\frac{d\phi_m(x)}{dx}dx = \frac{-1}{x_m - x_{m-1}}, \quad \text{for } i = m-1$$

$$\int_0^L \frac{d\phi_m(x)}{dx}\frac{d\phi_m(x)}{dx}dx = \frac{1}{x_m - x_{m-1}} + \frac{1}{x_{m+1} - x_m}, \quad \text{for } i = m$$

$$\int_0^L \frac{d\phi_m(x)}{dx}\frac{d\phi_{m+1}(x)}{dx}dx = \frac{-1}{x_{m+1} - x_m}, \quad \text{for } i = m+1$$

and, otherwise

$$\int_0^L \frac{d\phi_i(x)}{dx}\frac{d\phi_m(x)}{dx}dx = 0.$$

Thus, for a given $\phi_m(x)$ ($m \neq 0$ and $m \neq n$), only three terms in the last term of (3.36) remain and all other terms of the sum vanish. Then (3.36) will become;

$$-D\left\{\frac{1}{x_m - x_{m-1}}v_{m-1} - \left(\frac{1}{x_m - x_{m-1}} + \frac{1}{x_{m+1} - x_m}\right)v_m + \frac{1}{x_{m+1} - x_m}v_{m+1}\right\}$$

$$(3.37)$$

since $\phi_m(0) = 0$ and $\phi_m(L) = 0$ for $m \neq 0$ and $m \neq n$, and thus,

$$\frac{\partial \tilde{v}}{\partial x}(L,t)\phi_m(L) = 0$$

$$\frac{\partial \tilde{v}}{\partial x}(0,t)\phi_m(0) = 0$$

For $m = 0$, using $\phi_0(0) = 1$ and $\phi_0(L) = 0$, (3.36) becomes

$$D\frac{\partial \tilde{v}}{\partial x}(0,t) + \sum_{i=0}^{n} Dv_i \int_0^L \frac{d\phi_i(x)}{dx}\frac{d\phi_0(x)}{dx}dx$$

$$= D\frac{\partial v_0}{\partial x}(0,t) + Dv_0\frac{1}{x_1 - x_0} - Dv_1\frac{1}{x_1 - x_0}. \tag{3.38}$$

Similarly, for $m = n$, using $\phi_n(0) = 0$ and $\phi_n(L) = 1$, (3.36) becomes

$$D\frac{\partial \tilde{v}}{\partial x}(L,t) + \sum_{i=0}^{n} Dv_i \int_0^L \frac{d\phi_i(x)}{dx}\frac{d\phi_n(x)}{dx}dx$$

$$= -D\frac{\partial v_n}{\partial x}(L,t) - Dv_{n-1}\frac{1}{x_n - x_{n-1}} + Dv_n\frac{1}{x_n - x_{n-1}}. \tag{3.39}$$

The third term of (3.29) will be;

$$\int_0^L I(\tilde{v}, \tilde{w})\phi_m(x)dx$$

$$= \sum_{i=0}^{n}\sum_{j=0}^{n}\sum_{k=0}^{n} v_i(t)v_j(t)v_k(t) \int_0^L \phi_i(x)\phi_j(x)\phi_k(x)\phi_m(x)dx \tag{3.40}$$

$$- a\sum_{i=0}^{n}\sum_{k=0}^{n} v_i(t)v_k(t) \int_0^L \phi_i(x)\phi_k(x)\phi_m(x)dx \tag{3.41}$$

$$- b\sum_{i=0}^{n}\sum_{j=0}^{n} v_i(t)v_j(t) \int_0^L \phi_i(x)\phi_j(x)\phi_m(x)dx \tag{3.42}$$

$$+ ab\sum_{i=0}^{n} v_i(t) \int_0^L \phi_i(x)\phi_m(x)dx \tag{3.43}$$

$$+ \sum_{l=0}^{n} w_l(t) \int_0^L \phi_l(x)\phi_m(x)dx \tag{3.44}$$

For different combinations of the subscripts i, j, k, and m in (3.40),

$$\int_0^L \{\phi_m(x)\}^4\, dx = \frac{x_{m+1} - x_{m-1}}{5}, \quad \text{for } i = j = k = m$$

$$\int_0^L \{\phi_{m-1}(x)\}^2 \{\phi_m(x)\}^2 \, dx = \frac{x_m - x_{m-1}}{30}, \quad \text{for } i = j = m - 1 \text{ and } k = m$$

$$\text{or for } i = k = m - 1 \text{ and } j = m$$
$$\text{or for } j = k = m - 1 \text{ and } i = m$$

$$\int_0^L \{\phi_m(x)\}^2 \{\phi_{m+1}(x)\}^2 \, dx = \frac{x_{m+1} - x_m}{30}, \quad \text{for } i = j = m + 1 \text{ and } k = m$$

$$\text{or for } i = k = m + 1 \text{ and } j = m$$
$$\text{or for } j = k = m + 1 \text{ and } i = m$$

$$\int_0^L \{\phi_{m-1}(x)\}^3 \, \phi_m(x) dx = \frac{x_m - x_{m-1}}{20}, \quad \text{for } i = j = k = m - 1$$

$$\int_0^L \{\phi_m(x)\}^3 \, \phi_{m+1}(x) dx = \frac{x_{m+1} - x_m}{20}, \quad \text{for } i = j = m \text{ and } k = m + 1$$

$$\text{or for } i = k = m \text{ and } j = m + 1$$
$$\text{or for } j = k = m \text{ and } i = m + 1$$

$$\int_0^L \phi_{m-1}(x) \{\phi_m(x)\}^3 \, dx = \frac{x_m - x_{m-1}}{20}, \quad \text{for } i = m - 1 \text{ and } j = k = m$$

$$\text{or for } j = m - 1 \text{ and } i = k = m$$
$$\text{or for } k = m - 1 \text{ and } i = j = m$$

$$\int_0^L \phi_m(x) \{\phi_{m+1}(x)\}^3 \, dx = \frac{x_{m+1} - x_m}{20}, \quad \text{for } i = j = k = m + 1$$

and, otherwise

$$\int_0^L \phi_i(x)\phi_j(x)\phi_k(x)\phi_m(x) dx = 0.$$

Thus, for a given $\phi_m(x)$ ($m \neq 0$ and $m \neq n$), (3.40) becomes the sum of the following terms:

$$\frac{x_{m+1} - x_{m-1}}{5} v_m^3, \quad \text{for } i = j = k = m \tag{3.45}$$

$$\frac{x_m - x_{m-1}}{30} v_{m-1}^2 v_m, \quad \text{for } i = j = m - 1 \text{ and } k = m$$

$$\text{or for } i = k = m - 1 \text{ and } j = m$$
$$\text{or for } j = k = m - 1 \text{ and } i = m \tag{3.46}$$

$$\frac{x_{m+1} - x_m}{30} v_m v_{m+1}^2, \quad \text{for } i = j = m + 1 \text{ and } k = m$$

$$\text{or for } i = k = m + 1 \text{ and } j = m$$
$$\text{or for } j = k = m + 1 \text{ and } i = m \tag{3.47}$$

$$\frac{x_m - x_{m-1}}{20} v_{m-1}^3, \quad \text{for } i = j = k = m - 1 \tag{3.48}$$

$$\frac{x_{m+1} - x_m}{20} v_m^2 v_{m+1}, \quad \text{for } i = j = m \text{ and } k = m + 1$$

$$\text{or for } i = k = m \text{ and } j = m + 1$$

$$\text{or for } j = k = m \text{ and } i = m + 1 \tag{3.49}$$

$$\frac{x_m - x_{m-1}}{20} v_{m-1} v_m^2, \quad \text{for } i = m - 1 \text{ and } j = k = m$$

$$\text{or for } j = m - 1 \text{ and } i = k = m$$

$$\text{or for } k = m - 1 \text{ and } i = j = m \tag{3.50}$$

$$\frac{x_{m+1} - x_m}{20} v_{m+1}^3, \quad \text{for } i = j = k = m + 1 \tag{3.51}$$

and all other terms vanish.

We continue similar calculations for (3.41). For different combinations of the subscripts i, k, and m in (3.41),

$$\int_0^L \{\phi_m(x)\}^3 \, dx = \frac{x_{m+1} - x_{m-1}}{4}, \quad \text{for } i = k = m$$

$$\int_0^L \phi_{m-1}(x) \{\phi_m(x)\}^2 \, dx = \frac{x_m - x_{m-1}}{12}, \quad \text{for } i = m - 1 \text{ and } k = m$$

$$\text{or for } i = m \text{ and } k = m - 1$$

$$\int_0^L \phi_m(x) \{\phi_{m+1}(x)\}^2 \, dx = \frac{x_{m+1} - x_m}{12}, \quad \text{for } i = k = m + 1$$

$$\int_0^L \{\phi_{m-1}(x)\}^2 \phi_m(x) dx = \frac{x_m - x_{m-1}}{12}, \quad \text{for } i = k = m - 1$$

$$\int_0^L \{\phi_m(x)\}^2 \phi_{m+1}(x) dx = \frac{x_{m+1} - x_m}{12}, \quad \text{for } i = m + 1 \text{ and } k = m$$

$$\text{or for } k = m + 1 \text{ and } i = m$$

and, otherwise

$$\int_0^L \phi_i(x)\phi_k(x)\phi_m(x) dx = 0.$$

Thus, for a given $\phi_m(x)$ ($m \neq 0$ and $m \neq n$), (3.41) becomes the sum of the following terms:

$$-a\frac{x_{m+1} - x_{m-1}}{4} v_m^2, \quad \text{for } i = k = m \tag{3.52}$$

$$-a\frac{x_m - x_{m-1}}{12} v_{m-1} v_m, \quad \text{for } i = m - 1 \text{ and } k = m$$

$$\text{or for } i = m \text{ and } k = m - 1 \tag{3.53}$$

$$-a\frac{x_{m+1} - x_m}{12} v_{m+1}^2, \quad \text{for } i = k = m + 1 \tag{3.54}$$

$$-a\frac{x_m - x_{m-1}}{12}v_{m-1}^2, \quad \text{for } i = k = m - 1 \tag{3.55}$$

$$-a\frac{x_{m+1} - x_m}{12}v_m v_{m+1}, \quad \text{for } i = m + 1 \text{ and } k = m$$

$$\text{or for } k = m + 1 \text{ and } i = m \tag{3.56}$$

Using the same integra appeared in (3.42) for different combinations of the subscripts i, j, and m, for a given $\phi_m(x)$ ($m \neq 0$ and $m \neq n$), (3.42) becomes the sum of the following terms:

$$-b\frac{x_{m+1} - x_{m-1}}{4}v_m^2, \quad \text{for } i = j = m \tag{3.57}$$

$$-b\frac{x_m - x_{m-1}}{12}v_{m-1}v_m, \quad \text{for } i = m - 1 \text{ and } j = m$$

$$\text{or for } i = m \text{ and } j = m - 1 \tag{3.58}$$

$$-b\frac{x_{m+1} - x_m}{12}v_{m+1}^2, \quad \text{for } i = j = m + 1 \tag{3.59}$$

$$-b\frac{x_m - x_{m-1}}{12}v_{m-1}^2, \quad \text{for } i = j = m - 1 \tag{3.60}$$

$$-b\frac{x_{m+1} - x_m}{12}v_m v_{m+1}, \quad \text{for } i = m + 1 \text{ and } j = m$$

$$\text{or for } j = m + 1 \text{ and } i = m \tag{3.61}$$

Finally, a set of integra in (3.33) is used for simplifying (3.43) and (3.44) for a given $\phi_m(x)$ ($m \neq 0$ and $m \neq n$). Then (3.43) becomes the sum of the following terms:

$$ab\frac{x_m - x_{m-1}}{6}v_{m-1}, \quad \text{for } i = m - 1 \tag{3.62}$$

$$ab\frac{x_{m+1} - x_{m-1}}{3}v_m, \quad \text{for } i = m \tag{3.63}$$

$$ab\frac{x_{m+1} - x_m}{6}v_{m+1}, \quad \text{for } i = m + 1 \tag{3.64}$$

and, (3.44) becomes the sum of the following terms:

$$\frac{x_m - x_{m-1}}{6}w_{m-1}, \quad \text{for } l = m - 1 \tag{3.65}$$

$$\frac{x_{m+1} - x_{m-1}}{3}w_m, \quad \text{for } l = m \tag{3.66}$$

$$\frac{x_{m+1} - x_m}{6}w_{m+1}, \quad \text{for } l = m + 1 \tag{3.67}$$

If we assume equally spaced discretization of x for simplicity, i.e., $x_i - x_{i-1} \equiv \Delta x$ for all $i = 1, 2, \cdots, n - 1$, (3.29) can then be reexpressed as;

$$\frac{1}{6}\left(\frac{\partial v_{m-1}}{\partial t} + 4\frac{\partial v_m}{\partial t} + \frac{\partial v_{m+1}}{\partial t}\right) \tag{3.68}$$

$$-\frac{D}{\Delta x^2}\left(v_{m-1} - 2v_m + v_{m+1}\right) \tag{3.69}$$

$$+ \bar{i}_{ion}\left(v_{m-1}, v_m, v_{m+1}, w_{m-1}, w_m, w_{m+1}\right) \tag{3.70}$$

$$= 0$$

where

$$\bar{i}_{ion}(v_{m-1}, v_m, v_{m+1}, w_{m-1}, w_m, w_{m+1})$$

$$= \frac{1}{20}\left(8v_m^3 + 2v_{m-1}^2 v_m + 2v_m v_{m+1}^2 + v_{m-1}^3 + 3v_m^2 v_{m+1} + 3v_{m-1}v_m^2 + v_{m+1}^3\right)$$

$$-\frac{(a+b)}{12}\left(6v_m^2 + 2v_{m-1}v_m + v_{m+1}^2 + v_{m-1}^2 + 2v_m v_{m+1}\right)$$

$$+\frac{ab}{6}\left(v_{m-1} + 4v_m + v_{m+1}\right) + \frac{1}{6}\left(w_{m-1} + 4w_m + w_{m+1}\right) \tag{3.71}$$

for $m \neq 0$ and $m \neq n$. It is worthwhile to notice that the equation (the sum of (3.68), (3.69), and (3.70) is equal to zero) corresponds to (or is the same as) (3.5) if we put $v_m = v_{m-1} = v_{m+1}$ and $w_m = w_{m-1} = w_{m+1}$ in (3.68) and (3.70).

The third term of (3.29) for $m = 0$ becomes

$$\frac{x_1 - x_0}{5}v_0^3 + \frac{x_1 - x_0}{10}v_0 v_1^2 + \frac{3(x_1 - x_0)}{20}v_0^2 v_1 + \frac{x_1 - x_0}{20}v_1^3$$

$$-(a+b)\left(\frac{x_1 - x_0}{4}v_0^2 + \frac{x_1 - x_0}{12}v_1^2 + \frac{x_1 - x_0}{6}v_0 v_1\right)$$

$$+ab\left(\frac{x_1 - x_0}{3}v_0 + \frac{x_1 - x_0}{6}v_1\right) + \left(\frac{x_1 - x_0}{3}w_0 + \frac{x_1 - x_0}{6}w_1\right) \tag{3.72}$$

Assuming $x_1 - x_0 \equiv \Delta x$ also here, we have a representation of (3.29) for $m = 0$ as;

$$\frac{1}{6}\left(2\frac{\partial v_0}{\partial t} + \frac{\partial v_1}{\partial t}\right) + \frac{D}{\Delta x}\frac{\partial v_0}{\partial x} - \frac{D}{\Delta x^2}\left(v_1 - v_0\right)$$

$$+ \frac{1}{20}\left(4v_0^3 + 2v_0 v_1^2 + 3v_0^2 v_1 + v_1^3\right) - \frac{(a+b)}{12}\left(3v_0^2 + v_1^2 + 2v_0 v_1\right)$$

$$+ \frac{ab}{6}\left(2v_0 + v_1\right) + \frac{1}{6}\left(2w_0 + w_1\right) = 0 \tag{3.73}$$

In a similar way, for $m = n$ with $x_n - x_{n-1} \equiv \Delta x$, (3.29) becomes as follows:

$$\frac{1}{6}\left(\frac{\partial v_{n-1}}{\partial t} + 2\frac{\partial v_n}{\partial t}\right) - \frac{D}{\Delta x}\frac{\partial v_n}{\partial x} + \frac{D}{\Delta x^2}(v_n - v_{n-1})$$

$$+ \frac{1}{20}\left(4v_n^3 + 2v_{n-1}^2 v_n + 3v_{n-1}v_n^2 + v_{n-1}^3\right)$$

$$- \frac{(a+b)}{12}\left(3v_n^2 + v_{n-1}^2 + 2v_{n-1}v_n\right)$$

$$+ \frac{ab}{6}(2v_{n-1} + v_n) + \frac{1}{6}(2w_{n-1} + w_n) = 0 \quad (3.74)$$

We perform the same procedure for the differential equation

$$\frac{\partial w(x,t)}{\partial t} - \epsilon\left(v(x,t) - cw(x,t)\right) = 0 \quad (3.75)$$

by considering the following integral to become zero for every function $\phi_m(x)$ ($m = 1, 2, \ldots, n$):

$$\int_0^L \left[\frac{\partial \tilde{w}(x,t)}{\partial t} - \epsilon\left(\tilde{v}(x,t) - c\tilde{w}(x,t)\right)\right]\phi_m(x)dx = 0. \quad (3.76)$$

This is relatively easy, since the differential equation for $\tilde{w}$ does not include any nonlinear terms. For $m \neq 0$ and $m \neq n$, this becomes

$$\frac{1}{6}\left(\frac{\partial w_{m-1}}{\partial t} + 4\frac{\partial w_m}{\partial t} + \frac{\partial w_{m+1}}{\partial t}\right)$$

$$-\epsilon\left(\frac{1}{6}(v_{m-1} + 4v_m + v_{m+1}) - \frac{c}{6}(w_{m-1} + 4w_m + w_{m+1})\right) = 0.$$

Moreover, for $m = 0$,

$$\frac{1}{6}\left(2\frac{\partial v_0}{\partial t} + \frac{\partial v_1}{\partial t}\right) - \epsilon\left(\frac{1}{6}(2v_0 + v_1) - \frac{c}{6}(2w_0 + w_1)\right) = 0,$$

and, for $m = n$

$$\frac{1}{6}\left(\frac{\partial v_{n-1}}{\partial t} + 2\frac{\partial v_n}{\partial t}\right) - \epsilon\left(\frac{1}{6}(2v_{n-1} + v_n) - \frac{c}{6}(2w_{n-1} + w_n)\right) = 0.$$

In summary, by denoting

$$\mathbf{u} = (v_0, v_1, \cdots, v_{m-1}, v_m, v_{m+1}, \cdots, v_{n-1}, v_n,$$

$$w_0, w_1, \cdots, w_{m-1}, w_m, w_{m+1}, \cdots, w_{n-1}, w_n)^T,$$

and

$$\frac{\partial \mathbf{u}}{\partial t} = \left(\frac{\partial v_0}{\partial t}, \frac{\partial v_1}{\partial t}, \frac{\partial v_2}{\partial t}, \cdots, \frac{\partial v_{m-1}}{\partial t}, \frac{\partial v_m}{\partial t}, \frac{\partial v_{m+1}}{\partial t}, \cdots, \frac{\partial v_n}{\partial t}, \right.$$

$$\left. \frac{\partial w_0}{\partial t}, \frac{\partial w_1}{\partial t}, \frac{\partial w_2}{\partial t}, \cdots, \frac{\partial w_{m-1}}{\partial t}, \frac{\partial w_m}{\partial t}, \frac{\partial w_{m+1}}{\partial t}, \cdots, \frac{\partial w_n}{\partial t} \right)^T$$

Equations (3.29) and (3.30) can be written in a matrix form as;

$$\begin{pmatrix} A_v & 0 \\ 0 & A_w \end{pmatrix} \frac{\partial \mathbf{u}}{\partial t} - \begin{pmatrix} D_{vv} & 0 \\ 0 & 0 \end{pmatrix} \mathbf{u} + \begin{pmatrix} L_{vv} & L_{vw} \\ L_{wv} & L_{ww} \end{pmatrix} \mathbf{u} + I^{nl}(\mathbf{u}) + B = 0 \qquad (3.77)$$

Each of the five terms in (3.77) corresponds, in this order, to each of the five terms in the following continuous formulations rearranging (3.29) and (3.30):

$$\int_0^L \begin{pmatrix} \frac{\partial \tilde{v}}{\partial t} \\ \frac{\partial \tilde{w}}{\partial t} \end{pmatrix} \phi_m(x) dx - \left[- \int_0^L \begin{pmatrix} D \frac{\partial \tilde{v}}{\partial x} \\ 0 \end{pmatrix} \frac{d\phi_m(x)}{dx} dx \right]$$

$$+ \int_0^L \begin{pmatrix} I^{lin}(\tilde{v}, \tilde{w}) \\ \tilde{v} - c\tilde{w} \end{pmatrix} \phi_m(x) dx + \int_0^L \begin{pmatrix} I^{nl}(\tilde{v}, \tilde{w}) \\ 0 \end{pmatrix} \phi_m(x) dx$$

$$+ \begin{pmatrix} \left[D \frac{\partial \tilde{v}}{\partial x} \phi_m(x) \right]_L^0 \\ 0 \end{pmatrix} = 0 \qquad (3.78)$$

where

$$I(\tilde{v}, \tilde{w}) = I^{lin}(\tilde{v}, \tilde{w}) + I^{nl}(\tilde{v}, \tilde{w})$$

with the linear terms of $I(\tilde{v}, \tilde{w})$ defined as $I^{lin}(\tilde{v}, \tilde{w})$ and the nonlinear terms of $I(\tilde{v}, \tilde{w})$ defined as $I^{nl}(\tilde{v}, \tilde{w})$.

The matrices in (3.77) are defined as follows.

$$A_v = A_w = \begin{pmatrix} \frac{1}{3} & \frac{1}{6} & 0 & \cdots\cdots\cdots\cdots\cdots\cdots & 0 \\ \frac{1}{6} & \frac{2}{3} & \frac{1}{6} & 0 & \cdots\cdots\cdots\cdots\cdots & 0 \\ 0 & \frac{1}{6} & \frac{2}{3} & \frac{1}{6} & 0 & \cdots\cdots\cdots\cdots & 0 \\ \vdots & \vdots & \vdots & \vdots & \vdots & \vdots & \vdots & \vdots & \vdots & \vdots & \vdots \\ 0 & \cdots & 0 & \frac{1}{6} & \frac{2}{3} & \frac{1}{6} & 0 & \cdots\cdots & 0 \\ 0 & \cdots\cdots & 0 & \frac{1}{6} & \frac{2}{3} & \frac{1}{6} & 0 & \cdots & 0 \\ 0 & \cdots\cdots & 0 & \frac{1}{6} & \frac{2}{3} & \frac{1}{6} & 0 & \cdots & 0 \\ \vdots & \vdots & \vdots & \vdots & \vdots & \vdots & \vdots & \vdots & \vdots \\ 0 & \cdots\cdots\cdots\cdots\cdots & 0 & \frac{1}{6} & \frac{2}{3} & \frac{1}{6} \\ 0 & \cdots\cdots\cdots\cdots\cdots\cdots & 0 & \frac{1}{6} & \frac{1}{3} \end{pmatrix}, \qquad (3.79)$$

$$D_{vv} = \begin{pmatrix} D_{vv}^{0,0} & D_{vv}^{0,1} & 0 & \cdots & \cdots & \cdots & \cdots & \cdots & \cdots & \cdots & 0 \\ D_{vv}^{adj} & D_{vv}^{mid} & D_{vv}^{adj} & 0 & \cdots & \cdots & \cdots & \cdots & \cdots & \cdots & 0 \\ 0 & D_{vv}^{adj} & D_{vv}^{mid} & D_{vv}^{adj} & 0 & \cdots & \cdots & \cdots & \cdots & \cdots & 0 \\ \vdots & \vdots & \vdots & \vdots & \vdots & \vdots & \vdots & \vdots & \vdots & \vdots & \vdots \\ 0 & \cdots & \cdots & 0 & D_{vv}^{adj} & D_{vv}^{mid} & D_{vv}^{adj} & 0 & \cdots & \cdots & 0 \\ 0 & \cdots & \cdots & \cdots & 0 & D_{vv}^{adj} & D_{vv}^{mid} & D_{vv}^{adj} & 0 & \cdots & 0 \\ 0 & \cdots & \cdots & \cdots & \cdots & 0 & D_{vv}^{adj} & D_{vv}^{mid} & D_{vv}^{adj} & 0 & 0 \\ \vdots & \vdots & \vdots & \vdots & \vdots & \vdots & \vdots & \vdots & \vdots & \vdots & \vdots \\ 0 & \cdots & \cdots & \cdots & \cdots & \cdots & \cdots & 0 & D_{vv}^{adj} & D_{vv}^{mid} & D_{vv}^{adj} \\ 0 & \cdots & \cdots & \cdots & \cdots & \cdots & \cdots & \cdots & 0 & D_{vv}^{n,n-1} & D_{vv}^{n,n} \end{pmatrix}, \quad (3.80)$$

with

$$D_{vv}^{0,0} = -\frac{D}{\Delta x^2}, \quad D_{vv}^{0,1} = \frac{D}{\Delta x^2},$$

$$D_{vv}^{adj} = \frac{D}{\Delta x^2}, \quad D_{vv}^{mid} = -\frac{2D}{\Delta x^2},$$

$$D_{vv}^{n,n-1} = \frac{D}{\Delta x^2}, \quad D_{vv}^{n,n} = -\frac{D}{\Delta x^2},$$

$$L_{vv} = \begin{pmatrix} L_{vv}^{0,0} & L_{vv}^{0,1} & 0 & \cdots & \cdots & \cdots & \cdots & \cdots & \cdots & \cdots & 0 \\ L_{vv}^{adj} & L_{vv}^{mid} & L_{vv}^{adj} & 0 & \cdots & \cdots & \cdots & \cdots & \cdots & \cdots & 0 \\ 0 & L_{vv}^{adj} & L_{vv}^{mid} & L_{vv}^{adj} & 0 & \cdots & \cdots & \cdots & \cdots & \cdots & 0 \\ \vdots & \vdots & \vdots & \vdots & \vdots & \vdots & \vdots & \vdots & \vdots & \vdots & \vdots \\ 0 & \cdots & \cdots & 0 & L_{vv}^{adj} & L_{vv}^{mid} & L_{vv}^{adj} & 0 & \cdots & \cdots & 0 \\ 0 & \cdots & \cdots & \cdots & 0 & L_{vv}^{adj} & L_{vv}^{mid} & L_{vv}^{adj} & 0 & \cdots & 0 \\ 0 & \cdots & \cdots & \cdots & \cdots & 0 & L_{vv}^{adj} & L_{vv}^{mid} & L_{vv}^{adj} & 0 & 0 \\ \vdots & \vdots & \vdots & \vdots & \vdots & \vdots & \vdots & \vdots & \vdots & \vdots & \vdots \\ 0 & \cdots & \cdots & \cdots & \cdots & \cdots & \cdots & 0 & L_{vv}^{adj} & L_{vv}^{mid} & L_{vv}^{adj} \\ 0 & \cdots & \cdots & \cdots & \cdots & \cdots & \cdots & \cdots & 0 & L_{vv}^{n,n-1} & L_{vv}^{n,n} \end{pmatrix}, \quad (3.81)$$

with

$$L_{vv}^{0,0} = \frac{ab}{3}, \quad L_{vv}^{0,1} = \frac{ab}{6},$$

$$L_{vv}^{adj} = \frac{ab}{6}, \quad L_{vv}^{mid} = \frac{2ab}{3},$$

$$L_{vv}^{n,n-1} = \frac{ab}{3}, \quad L_{vv}^{n,n} = \frac{ab}{6},$$

$$L_{ww} = \begin{pmatrix} L_{ww}^{0,0} & L_{ww}^{0,1} & 0 & \cdots & \cdots & \cdots & \cdots & \cdots & \cdots & \cdots & \cdots & 0 \\ L_{ww}^{adj} & L_{ww}^{mid} & L_{ww}^{adj} & 0 & \cdots & \cdots & \cdots & \cdots & \cdots & \cdots & \cdots & 0 \\ 0 & L_{ww}^{adj} & L_{ww}^{mid} & L_{ww}^{adj} & 0 & \cdots & \cdots & \cdots & \cdots & \cdots & \cdots & 0 \\ \vdots & \vdots & \vdots & \vdots & \vdots & \vdots & \vdots & \vdots & \vdots & \vdots & \vdots & \vdots \\ 0 & \cdots & \cdots & 0 & L_{ww}^{adj} & L_{ww}^{mid} & L_{ww}^{adj} & 0 & \cdots & \cdots & \cdots & 0 \\ 0 & \cdots & \cdots & \cdots & 0 & L_{ww}^{adj} & L_{ww}^{mid} & L_{ww}^{adj} & 0 & \cdots & \cdots & 0 \\ 0 & \cdots & \cdots & \cdots & \cdots & 0 & L_{ww}^{adj} & L_{ww}^{mid} & L_{ww}^{adj} & 0 & \cdots & 0 \\ \vdots & \vdots & \vdots & \vdots & \vdots & \vdots & \vdots & \vdots & \vdots & \vdots & \vdots & \vdots \\ 0 & \cdots & \cdots & \cdots & \cdots & \cdots & \cdots & \cdots & 0 & L_{ww}^{adj} & L_{ww}^{mid} & L_{ww}^{adj} \\ 0 & \cdots & \cdots & \cdots & \cdots & \cdots & \cdots & \cdots & 0 & L_{ww}^{n,n-1} & L_{ww}^{n,n} \end{pmatrix}, \quad (3.82)$$

with

$$L_{ww}^{0,0} = \frac{c\epsilon}{3}, \quad L_{ww}^{0,1} = \frac{c\epsilon}{6},$$

$$L_{ww}^{adj} = \frac{c\epsilon}{6}, \quad L_{ww}^{mid} = \frac{2c\epsilon}{3},$$

$$L_{ww}^{n,n-1} = \frac{c\epsilon}{3}, \quad L_{ww}^{n,n} = \frac{c\epsilon}{6},$$

$$L_{vw} = \begin{pmatrix} L_{vw}^{0,0} & L_{vw}^{0,1} & 0 & \cdots & \cdots & \cdots & \cdots & \cdots & \cdots & \cdots & \cdots & 0 \\ L_{vw}^{adj} & L_{vw}^{mid} & L_{vw}^{adj} & 0 & \cdots & \cdots & \cdots & \cdots & \cdots & \cdots & \cdots & 0 \\ 0 & L_{vw}^{adj} & L_{vw}^{mid} & L_{vw}^{adj} & 0 & \cdots & \cdots & \cdots & \cdots & \cdots & \cdots & 0 \\ \vdots & \vdots & \vdots & \vdots & \vdots & \vdots & \vdots & \vdots & \vdots & \vdots & \vdots & \vdots \\ 0 & \cdots & \cdots & 0 & L_{vw}^{adj} & L_{vw}^{mid} & L_{vw}^{adj} & 0 & \cdots & \cdots & \cdots & 0 \\ 0 & \cdots & \cdots & \cdots & 0 & L_{vw}^{adj} & L_{vw}^{mid} & L_{vw}^{adj} & 0 & \cdots & \cdots & 0 \\ 0 & \cdots & \cdots & \cdots & \cdots & 0 & L_{vw}^{adj} & L_{vw}^{mid} & L_{vw}^{adj} & 0 & \cdots & 0 \\ \vdots & \vdots & \vdots & \vdots & \vdots & \vdots & \vdots & \vdots & \vdots & \vdots & \vdots & \vdots \\ 0 & \cdots & \cdots & \cdots & \cdots & \cdots & \cdots & \cdots & 0 & L_{vw}^{adj} & L_{vw}^{mid} & L_{vw}^{adj} \\ 0 & \cdots & \cdots & \cdots & \cdots & \cdots & \cdots & \cdots & 0 & L_{vw}^{n,n-1} & L_{vw}^{n,n} \end{pmatrix}, \quad (3.83)$$

with

$$L_{vw}^{0,0} = \frac{1}{3}, \quad L_{vw}^{0,1} = \frac{1}{6},$$

$$L_{vw}^{adj} = \frac{1}{6}, \quad L_{vw}^{mid} = \frac{2}{3},$$

$$L_{vw}^{n,n-1} = \frac{1}{3}, \quad L_{vw}^{n,n} = \frac{1}{6},$$

$$
L_{wv} = \begin{pmatrix}
L_{wv}^{0,0} & L_{wv}^{0,1} & 0 & \cdots & \cdots & \cdots & \cdots & \cdots & \cdots & \cdots & \cdots & 0 \\
L_{wv}^{adj} & L_{wv}^{mid} & L_{wv}^{adj} & 0 & \cdots & \cdots & \cdots & \cdots & \cdots & \cdots & \cdots & 0 \\
0 & L_{wv}^{adj} & L_{wv}^{mid} & L_{wv}^{adj} & 0 & \cdots & \cdots & \cdots & \cdots & \cdots & \cdots & 0 \\
\vdots & \vdots & \vdots & \vdots & \vdots & \vdots & \vdots & \vdots & \vdots & \vdots & \vdots & \vdots \\
0 & \cdots & \cdots & 0 & L_{wv}^{adj} & L_{wv}^{mid} & L_{wv}^{adj} & 0 & \cdots & \cdots & \cdots & 0 \\
0 & \cdots & \cdots & \cdots & 0 & L_{wv}^{adj} & L_{wv}^{mid} & L_{wv}^{adj} & 0 & \cdots & \cdots & 0 \\
0 & \cdots & \cdots & \cdots & \cdots & 0 & L_{wv}^{adj} & L_{wv}^{mid} & L_{wv}^{adj} & 0 & \cdots & 0 \\
\vdots & \vdots & \vdots & \vdots & \vdots & \vdots & \vdots & \vdots & \vdots & \vdots & \vdots & \vdots \\
0 & \cdots & \cdots & \cdots & \cdots & \cdots & \cdots & 0 & L_{wv}^{adj} & L_{wv}^{mid} & L_{wv}^{adj} \\
0 & \cdots & \cdots & \cdots & \cdots & \cdots & \cdots & \cdots & 0 & L_{wv}^{n,n-1} & L_{wv}^{n,n}
\end{pmatrix}, \quad (3.84)
$$

with

$$
L_{wv}^{0,0} = -\tfrac{\epsilon}{3}, \quad L_{wv}^{0,1} = -\frac{\epsilon}{6},
$$

$$
L_{wv}^{adj} = -\tfrac{\epsilon}{6}, \quad L_{wv}^{mid} = -\frac{2\epsilon}{3},
$$

$$
L_{wv}^{n,n-1} = -\tfrac{\epsilon}{3}, \quad L_{wv}^{n,n} = -\frac{\epsilon}{6},
$$

and

$$
I^{nl}(\mathbf{u}) = \begin{pmatrix}
I_0^{nl}(v_0, v_1) \\
\vdots \\
I_m^{nl}(v_{m-1}, v_m, v_{m+1}) \\
\vdots \\
I_n^{nl}(v_{n-1}, v_n) \\
0 \\
\vdots \\
0
\end{pmatrix} \quad (3.85)
$$

where $I^{nl}(\mathbf{u})$ is the $(n+1) \times 2$ vector for which I_0^{nl} is defined by the nonlinear terms (the first and second rows) of (3.72), I_m^{nl} by the nonlinear terms (the first and second rows) of (3.71), and I_n^{nl} by the nonlinear terms (the second and third rows) of (3.74), and the last half $(n+1)$ components are all zeros. $I_0^{nl}, \cdots, I_m^{nl}, \cdots, I_n^{nl}$ do not include $w_0, \cdots, w_n$ because $w(x,t)$ appears only as linear terms in the FHN.

The last term of (3.77), B, is also the $(n+1) \times 2$ vector, and it is defined as follows:

$$
B = \begin{pmatrix} B_0 \\ B_1 \\ \vdots \\ B_{n-1} \\ B_n \\ B_{n+1} \\ \vdots \\ B_{2(n+1)} \end{pmatrix} = \begin{pmatrix} \dfrac{D}{\Delta x}\dfrac{\partial v_0}{\partial x} \\ 0 \\ \vdots \\ 0 \\ -\dfrac{D}{\Delta x}\dfrac{\partial v_n}{\partial x} \\ 0 \\ \vdots \\ 0 \end{pmatrix} \tag{3.86}
$$

in which the last half $(n+1)$ components are all zeros.

Let us solve the partial differential equation (3.3) with Neumann boundary conditions at both terminals of the cable (Case 1 defined above) such that

$$
\frac{\partial v}{\partial x}(0,t) = 0,
$$

$$
\frac{\partial v}{\partial x}(L,t) = 0
$$

which are equivalent with the setting

$$
\frac{\partial v_0}{\partial x} = 0,
$$

$$
\frac{\partial v_n}{\partial x} = 0
$$

in the ordinary differential equation (3.77) approximating (3.3). Each of these conditions is sometimes called "free" boundary because the dynamical values at the boundary can change freely in time. With these conditions, the last term of (3.77), B, vanishes, and thus (3.77) can be numerically integrated easily using, for example, either forward or backward Euler discretization with respect to the time t.

Figure 3.6 exemplifies a time evolution of excitation propagation along the cable of (3.3). It shows the excitation propagation that induced by the initial square-shaped local excitation of the cable close to the left terminal of the cable. The initial local excitation induces the action potentials conducting both the left and right direction of the cable. However, only the one propagating rightward remains since there is no cable in the left-side of the initial local excitation. The rightward conducting action potential eventually reaches the right terminal of the cable and vanishes. Notice that the spatial derivatives of the v variable at the left and right terminals of the cable are always zeros because of the Neumann boundary conditions, $\frac{\partial v}{\partial x}(0) = \frac{\partial v}{\partial x}(L) = 0$.

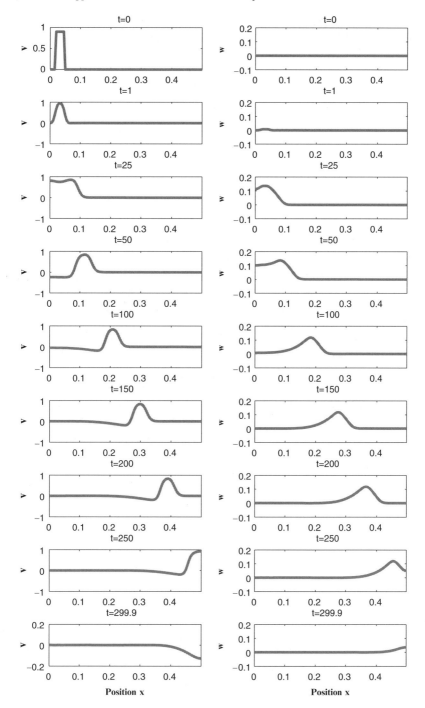

Fig. 3.6 Conduction of the action potential on one dimensional cable of FHN with $n = 100$, $D = 0.00002$, $\Delta x = 0.005$, $\Delta t = 0.1$, with "free" boundary conditions, i.e., $\frac{\partial v}{\partial x}(0) = \frac{\partial v}{\partial x}(L) = 0$ at left and right terminals of the cable, and with the initial condition at $t = 0$ having a local excitation of the cable close to the left terminal

IDE Modeling: FitzHugh–Nagumo Action Potential Conduction: Search ModelDB by "FHN Cable" to Find FHN_Cable_Neumann.isml

$$\frac{\partial v(x,t)}{\partial t} = D\frac{\partial^2 v(x,t)}{\partial x^2} - v(x,t)(v(x,t)-a)(v(x,t)-b) + w(x,t)$$

$$\frac{\partial w(x,t)}{\partial t} = \epsilon\,(v(x,t) - cw(x,t)) \tag{3.87}$$

with the boundary conditions

$$\frac{\partial v}{\partial x}(0,t) = 0,$$

$$\frac{\partial v}{\partial x}(L,t) = 0 \tag{3.88}$$

The *insilico*IDE can export a FreeFEM++[1] script for solving partial differential equations. FreeFEM++ is a software program developed by Universite Pierre et Marie Curie Laboratoire Jacques-Louis Lions. It is an implementation of a language dedicated to the finite element method, and enables you to solve Partial Differential Equations (PDE) easily. Figure 3.7 shows a snapshot of FreeFEM++ simulation of (3.87), on which the membrane potential v is color-coded to display spatio-temporal dynamics of the action potential conduction.

Let us now solve the same partial differential equation (3.3) but now with a Neumann and a Dirichlet boundary conditions, respectively, at the left ($x = 0$) and the right ($x = L$) terminals of the cable (Case 2 defined above) such that

$$\frac{\partial v}{\partial x}(0,t) = 0,$$
$$v(L,t) = \bar{v}$$

which are equivalent with the setting

$$\frac{\partial v_0}{\partial x} = 0,$$
$$v_n = \bar{v},$$

in (3.77). In this case, a numerical simulation of the model becomes slightly complicated. The coefficient v_n for the terminal node, which was a unknown variable that we needed to determine in the previous case, is now always fixed at a given constant value $\bar{v}$, and thus $\frac{\partial v_n}{\partial t} = 0$. Therefore, the dynamical variable that we need to determine during the numerical integration for Case 2 is

---

[1] http://www.freefem.org/ff++/.

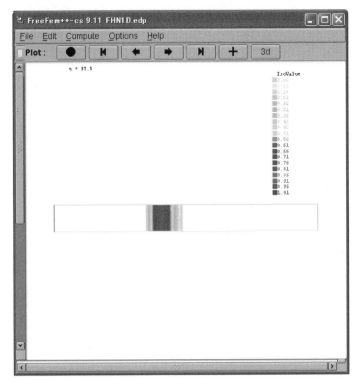

Fig. 3.7 A snapshot of *insilico*IDE for the action potential conduction in FHN model, and its simulation outcome

$$\mathbf{u}^* = (v_0, v_1, \cdots, v_{m-1}, v_m, v_{m+1}, \cdots, v_{n-1},$$
$$w_0, w_1, \cdots, w_{m-1}, w_m, w_{m+1}, \cdots, w_{n-1}, w_n)^T,$$

which is a $2(n+1) - 1$ dimensional vector instead of the $2(n+1)$ dimensional vector $\mathbf{u}$ for Case 1. Technically speaking, once we obtain $\mathbf{u}^*(t)$ at time t, we can calculate $\partial v_n / \partial x$ in B, the last term of (3.77) if it is desired.

Exercise 3.3. The reader is asked to write down every row component of (3.77), by which one can confirm that the $(n+1)$-th row here does not include any derivative with respect to time t, but only the spatial derivative of v_n.

Thus we have a set of $2(n+1) - 1$ ordinary differential equations. Let us define $n \times n$ matrices A_v^*, D_{vv}^*, and L_{vv}^*, respectively, as the sub-matrices of A_v, D_{vv}, and L_{vv} as follows;

$$A_v^* = \begin{pmatrix} A_v^{0,0} & \cdots & A_v^{0,n-1} \\ \vdots & \vdots \vdots & \vdots \\ A_v^{n-1,0} & \cdots & A_v^{n-1,n-1} \end{pmatrix},$$

$$D_{vv}^* = \begin{pmatrix} D_{vv}^{0,0} & \cdots & D_{vv}^{0,n-1} \\ \vdots & \vdots \vdots & \vdots \\ D_{vv}^{n-1,0} & \cdots & D_{vv}^{n-1,n-1} \end{pmatrix},$$

$$L_{vv}^* = \begin{pmatrix} L_{vv}^{0,0} & \cdots & L_{vv}^{0,n-1} \\ \vdots & \vdots \vdots & \vdots \\ L_{vv}^{n-1,0} & \cdots & L_{vv}^{n-1,n-1} \end{pmatrix},$$

and also L_{vw}^* of $n \times (n+1)$ matrix and L_{wv}^* of $(n+1) \times n$ matrix as;

$$L_{vw}^* = \begin{pmatrix} L_{vw}^{0,0} & \cdots & L_{vw}^{0,n} \\ \vdots & \vdots \vdots & \vdots \\ L_{vw}^{n-1,0} & \cdots & L_{vw}^{n-1,n} \end{pmatrix},$$

$$L_{wv}^* = \begin{pmatrix} L_{wv}^{0,0} & \cdots & L_{wv}^{0,n-1} \\ \vdots & \vdots \vdots & \vdots \\ L_{wv}^{n,0} & \cdots & L_{wv}^{n,n-1} \end{pmatrix}.$$

Then, the ordinary differential equation for $\mathbf{u}^*$ is represented as;

$$\begin{pmatrix} A_v^* & 0_{n\times(n+1)} \\ 0_{(n+1)\times n} & A_w \end{pmatrix} \frac{\partial \mathbf{u}^*}{\partial t} - \begin{pmatrix} D_{vv}^* & 0_{n\times(n+1)} \\ 0_{(n+1)\times n} & 0_{n\times n} \end{pmatrix} \mathbf{u}^*$$

$$+ \begin{pmatrix} L_{vv}^* & L_{vw}^* \\ L_{wv}^* & L_{ww} \end{pmatrix} \mathbf{u}^* + I^*(\mathbf{u}^*, \bar{v}) + C^*(\bar{v}) = 0 \qquad (3.89)$$

where I^* is the $[2(n+1)-1]$ vector defined as

$$I^*(\mathbf{u}^*, \bar{v}) = \begin{pmatrix} I_0(v_0, v_1) \\ \vdots \\ I_m(v_{m-1}, v_m, v_{m+1}) \\ \vdots \\ I_{n-1}(v_{n-2}, v_{n-1}, \bar{v}) \\ 0 \\ \vdots \\ 0 \end{pmatrix}, \qquad (3.90)$$

and $C^*(\bar{v})$ is the $[2(n+1)-1]$ vector defined as

$$
C^*(\bar{v}) = \begin{pmatrix} 0 \\ \vdots \\ 0 \\ (-D_{vv}^{n-1,n} + L_{vv}^{n-1,n})\bar{v} \leftarrow n\text{-th row} \\ 0 \\ \vdots \\ L_{wv}^{n-1,n}\bar{v} \leftarrow [2(n+1)-2]\text{-th row} \\ L_{wv}^{n,n}\bar{v} \leftarrow [2(n+1)-1]\text{-th row} \end{pmatrix}, \tag{3.91}
$$

The term C^* is associated with the product of $\mathbf{u}$ and the matrix elements that have been eliminated from D_{vv}, L_{vv}, and L_{wv} when we define the sub-matrices D_{vv}^*, L_{vv}^*, and L_{wv}^*.

The following equality, representing the $(n+1)$-th row of (3.77), is satisfied for any time instance t of the time evolution of the system defined by (3.89);

$$
A_{vv}^{n,n-1}v_{n-1} + A_{vv}^{n,n}\bar{v} - D_{vv}^{n,n-1}v_{n-1} - D_{vv}^{n,n}\bar{v}
$$

$$
+L_{vv}^{n,n-1}v_{n-1}(t) + L_{vv}^{n,n}\bar{v} + L_{vw}^{n,n-1}w_{n-1}(t) + L_{vw}^{n,n}w_n(t)
$$

$$
+I_n(v_{n-1}(t), \bar{v}, w_{n-1}(t), w_n(t)) + B_n\left(\frac{\partial v_n(t)}{\partial x}\right) = 0 \tag{3.92}
$$

by which the unknown derivative $\partial v_n(t)/\partial x$ at the right terminal of the cable can be determined if necessary.

As one can see in Fig. 3.8, the initial local excitation of the cable close to the left terminal of the cable conducts rightward as in the previous case. Moreover, the voltage clamp at the right terminal of the cable also induce the action potential that conduct the cable leftward. These two propagating excitation collide with each other at the middle of the cable, eventually leading to the annihilation of the conducting waves. Notice that the v value at the right terminal of the cable is always clamped at 0.5 due to the Dirichlet condition.

3.2.4 Excitation Conduction Through the Whole Heart

The heart is a muscular organ responsible for pumping blood throughout the blood vessels. Rhythmic and spatio-temporal coordinated muscle contractions make this function possible. The coordination is established by the propagation (conduction) of action potentials that are initiated, for every beat, at a small region of the heart, called *sinoatrial node* or SA-node, located at the right atrium in humans. The SA-node is an aggregation of spontaneously (autonomously) beating cells,

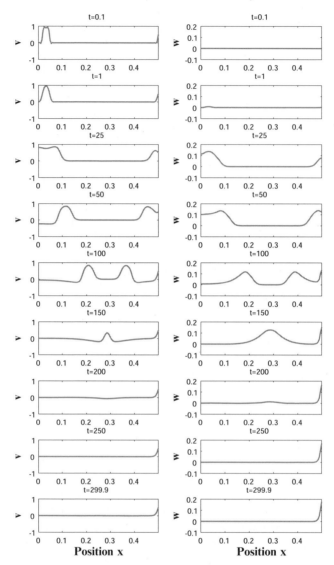

Fig. 3.8 Conduction of the action potential on one dimensional cable of FHN with $n = 100$, $D = 0.00002$, $\Delta x = 0.005$, $\Delta t = 0.1$ with "free" boundary condition at the left terminal, i.e., $\frac{\partial v}{\partial x}(0) = 0$ and the Dirichlet condition at the right terminal, i.e., $v(L) = 0.5$, and with the initial condition at $t = 0$ having a local excitation of the cable close to the left terminal

referred to as the *pacemaker cells*, and these cells generate action potentials almost synchronously. The action potential then conducts and spreads through the wall of the right and left atria. This is possible because cardiac muscle cells are electrically coupled by the gap junctions spanning between adjacent cells. The atria cells contract during the excitations of the cells. The action potential keeps conducting into a region of specialized cardiac muscle tissue connecting between the atrium and

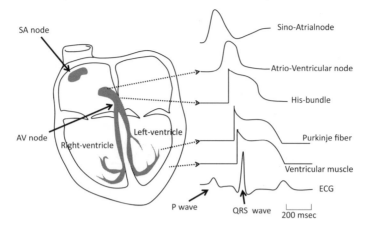

Fig. 3.9 Excitation conduction through whole heart and electrical morphology of cardiac action potentials

the ventricle, referred to as the *atrioventricular node* (AV-node). Due to the slow conduction characteristics of the AV-node, it usually takes about 0.1 s for the action potential to reach at the ventricle through the AV-node. This delay indeed ensures that the blood in the atria flows into the ventricles before the ventricles contract. A fiber bundle (His-bundle) at the AV-node continues to a network of specialized muscle fibers, referred to as the *Purkinje fibers*, in the wall of the ventricles. The action potentials conduct along this network and then throughout the ventricular walls. See Fig. 3.9.

There have been a large number of studies for simulating action potential conduction through the whole heart. See a book by Panfilov and Holden (1997) for example. Simulations of mechanical contractions of the heart with conducting action potentials and with the blood flows have also been studied. See Watanabe et al. (2004) for example.

3.3 Synaptically Connected Neuronal Networks

We have considered models of cellular networks in which cells communicate electrically via gap junctions. Another important way of communications among cellular membranes is established by *synapses*. Indeed, the gap junction is a type of synapses, and it is an electrical synapse. Here we consider another type of synapses, referred to as the *chemical synapse*. Figure 3.10 illustrates four neuronal cells that are inter-connected by chemical synapses. Each neuron is represented as its cell body called *soma* with a nucleus, dendrites (small tree-like components that come out from the cell body), and axon (a cable-like nerve fiber initiated from the soma). An axon can have several or hundreds of branches.

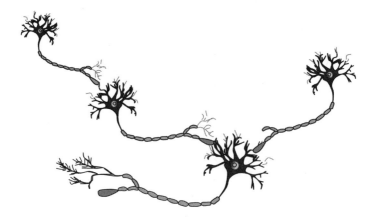

Fig. 3.10 A simple model of neuronal network

Let us consider two neurons. A chemical synapse is a junction formed between a terminal of the axon of one neuron referred to as the *presynaptic neuron* and the cell membrane (dendrites or cell body in many cases) of the other neuron referred to as the *postsynaptic neuron*. At the terminal, there exist vesicles which store chemicals called *neurotransmitters* as well as voltage-gated channels. Action potentials generated at the soma of the presynaptic neuron conduct along its axon and arrive at a synapse. Then the voltage-gated channels open, and neurotransmitters that are stored in the vesicles are released into a small space referred to as the *synaptic cleft* between the pre- and postsynaptic neurons. Those neurotransmitters bind to receptors on the postsynaptic neuron, triggering opening of synaptic channels that are located near the receptors on the postsynaptic neuron. Opening of the synaptic channels causes influx or efflux of ion currents as in the non-synaptic ion channels responsible for regulating the membrane potential.

There are many neurotransmitters such as glutamate and γ-aminobutyric (GABA). They are amino acids. Monoamines such as dopamine, norepinephrine, histamine, and serotonin are types of neurotransmitters. Others like acetylcholine, adenosine, and nitric oxide are also neurotransmitters. A specific type of neurotransmitters can bind a specific type of receptors on the postsynaptic membrane. Moreover, a specific type of receptor open a specific type of synaptic channels of the postsynaptic membrane. A combination of these three agents determine how the arrival of an action potential from the presynaptic neuron affects on the postsynaptic neuron via the synapse.

Synaptic influences can be either *excitatory* or *inhibitory*. Typically, glutamate transmitter is excitatory, and GABA is inhibitory, although, as mentioned, synaptic influences are determined by the combination of the transmitter, the receptor, and the channel. Let us focus the channel property. A synaptic current acts as excitatory or inhibitory depending on the reversal potential E_{syn} of ions carried through the synaptic channel. Unlike usual membrane channels, synaptic channels are typically permeable for several kinds of ions. Thus, E_{syn} may be determined by a combination of several Nernst potentials for several ion types such as for Na^+ and K^+. Direction of the synaptic current flux through the synaptic channel is then determined by the potential difference $V - E_{syn}$ where V is the postsynaptic membrane potential. Opening of the synaptic channel induces influx of the ions (cations) if $V - E_{syn} < 0$, leading to a negative inward current contributing to an increase in the membrane potential. For a specific synaptic channel, it is an excitatory if E_{syn} is high (mostly close to 0 mV and higher than the resting potential) and $V - E_{syn} < 0$ holds "most of the time." The use of the words "most of the time" is due to the fact that the membrane potential V changes during action potential, and thus $V - E_{syn}$ can be both negative and positive dynamically. That is, $V - E_{syn}$ can be positive for a short time interval even for excitatory synapses. A case if $V - E_{syn} > 0$ holds "most of the time", in which E_{syn} is lower than the resting potential, induces mostly efflux of the ions (a positive outward current by cations), and it corresponds to an inhibitory synapse.

In summary, a synaptic current I_{syn} can be modeled as

$$I_{syn} = g_{syn}(V - E_{syn}) \tag{3.93}$$

where g_{syn} represents the synaptic conductance, V the membrane potential, and E_{syn} the reversal potential of the ions that carry the synaptic current. g_{syn} may change dynamically, and its dynamics may also be modeled by ODEs. The synaptic current I_{syn} will appear in a dynamic equation of a membrane potential, together with a membrane ion channel current i_{ion} as

$$C\frac{dV}{dt} = -i_{ion} - I_{syn}.$$

If a synapse is excitatory, a presynaptic action potential causes a rise in the postsynaptic membrane potential, referred to as the *EPSP*, excitatory postsynaptic potential. If it is inhibitory, a presynaptic firing causes a drop in the postsynaptic membrane potential, referred to as the *IPSP*, inhibitory postsynaptic potential. The size (the amplitude) of the EPSP or IPSP is determined by the synaptic strength. The larger the synaptic strength, the larger the size of the EPSP or the IPSP. The synaptic strength is determined by the quantity of the presynaptic neurotransmitter release, the number of the postsynaptic receptors as well as the conductance of the synaptic channel. In many modeling, however, the overall synaptic strength is represented just by a value of g_{syn}.

Let us consider a response of a postsynaptic cellular membrane potential to a sequence of synaptic current injections using a very simple model that we considered before in (2.18). Here a single synaptic current generated by an arrival of single presynaptic action potential at the synapse at a time instant t is modeled simply by an ideal impulse $s(t) = -w\delta(t)$ where w represents the amplitude of the impulsive current $s(t)$, and $\delta(t)$ is the Dirac delta function (unit impulse function). By mathematical definition, the following equalities hold;

$$\int_{-\infty}^{\infty} \delta(t)dt = \int_{-\epsilon}^{+\epsilon} \delta(t)dt = 1$$

for any positive ϵ (even $\epsilon \to 0$), and

$$\int_{-\infty}^{\infty} f(t)\delta(t-a)dt = \int_{a-\epsilon}^{a+\epsilon} f(t)\delta(t-a)dt = f(a) \tag{3.94}$$

for any real number a and any smooth function $f(t)$.

Let us consider a case where the cell model receives a sequence of N impulses of $s(t)$, each of which is separated by the identical interspike time interval T generated by a presynaptic cell. For simplicity, we assume that the Nernst potential $E = 0$ in (2.18). Then we have the following model of the postsynaptic membrane potential.

$$C\frac{dV}{dt} = -\frac{1}{R}V - \sum_{n=0}^{N-1} s(t-nT), \tag{3.95}$$

where, as in (2.18), C represents the membrane capacitance, R the membrane resistance, V the membrane potential. Assuming that the cell model does not receive any synaptic input until $t = 0$, we have the initial condition at $t = 0$ as $V(0) = 0$ (the resting potential) for this ODE. The first synaptic current pulse arrives at $t = 0$. Since (3.95) is linear ODE with time-dependent external input, it can be solved analytically by various ways. The response of this cell model to a single $s(t)$ applied at $t = 0$ (the first input) can be obtained by solving

$$\frac{dV}{dt} = -\frac{1}{RC}V - \frac{1}{C}s(t). \tag{3.96}$$

We know that a solution of the corresponding homogeneous ODE

$$\frac{dV}{dt} = -\frac{1}{RC}V$$

is expressed as $V(t) = K\exp(-t/(RC))$ where K is a constant. Let us assume a solution of (3.96) with a form $V(t) = K(t)\exp(-t/(RC))$ with a time-dependent K, and determine $K(t)$ so that this $V(t)$ satisfies (3.96) indeed. Substituting this form into (3.96), we have the ODE for $K(t)$ as

$$\frac{dK}{dt} = -\frac{1}{C}\exp\left(\frac{t}{RC}\right)s(t). \tag{3.97}$$

Integrating both sides of this ODE from the initial time 0 to t, we have

$$K(t) = K(0) - \frac{1}{C} \int_0^t \exp\left(\frac{\tau}{RC}\right) s(\tau) d\tau, \tag{3.98}$$

and, thus

$$V(t) = V(0) \exp\left(-\frac{t}{RC}\right) - \frac{1}{C} \int_0^t \exp\left(-\frac{t-\tau}{RC}\right) s(\tau) d\tau. \tag{3.99}$$

using $K(0) = V(0)$ which is obtained from the fact that this solution $V(t)$ at $t = 0$ must be $V(0)$. Substituting $s(t) = -w\delta(t)$ and the initial condition $V(0) = 0$ into this solution, we have

$$V(t) = \frac{w}{C} \int_0^t \exp\left(-\frac{t-\tau}{RC}\right) \delta(\tau) d\tau. \tag{3.100}$$

Then, using (3.94), the solution of (3.96) is obtained as

$$V(t) = \frac{w}{C} \exp\left(-\frac{t}{RC}\right). \tag{3.101}$$

In the dynamic system theory, this solution is referred to as the *impulse response function* of the system. Now, because of the linearity of the ODE in (3.95), the membrane potential response to the sequence of the impulsive synaptic currents can be represented by the sum of the impulse response functions for single impulse synaptic current injections at $t = 0, t = T, t = 2T, \ldots$, and $t = (N-1)T$ as

$$V(t) = \frac{w}{C} \sum_{n=0}^{N-1} \exp\left(-\frac{t-nT}{RC}\right). \tag{3.102}$$

A solution in (3.102) is exemplified in Fig. 3.11, in which we set the period $T = 5\,\mathrm{ms}$, the intensity $w = 1.0$, and the time constant $RC = 10\,\mathrm{ms}$. The first synaptic current is injected at $t = 0$, and in total, ten synaptic inputs are applied. As one can see in the figure, the membrane potential of the model is shifted up by each synaptic current, reproducing its corresponding EPSP. After the rapid increase, the membrane potential decays toward zero with the time constant RC. In this case, however, EPSPs are piled up and integrated to reach about 2.5 mV since the inter-synaptic current intervals are smaller than the time constant, and every subsequent current is injected before the membrane potential returns to zero. After the termination of the input sequence, the membrane potential decays to zero with the time constant RC.

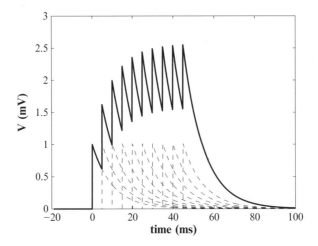

Fig. 3.11 Synapse integrations in a simple RC-circuit cell model. *Thick curve*: Membrane potential $V(t)$ in response to a sequence of ten impulsive synaptic currents. *Dashed curves*: Each of these is a membrane potential evoked in response to a single impulsive synaptic current

In (3.95), the synaptic current is simply modeled by the impulse function $\delta(t)$ without use of (3.93). Here we consider more realistic but still simple model of synaptic currents. Let us consider a single postsynaptic neuron receiving N synapses from some presynaptic neurons. Dynamics of the n-th synapse conductance $g_{syn}^{(n)}$ of the postsynaptic neuron are modeled by a simple first order ODE as

$$\frac{dg_{syn}^{(n)}}{dt} = -\frac{g_{syn}^{(n)}}{\tau_{syn}^{(n)}} + \sum_{i(n)} w_n \delta(t - t_{i(n)}). \tag{3.103}$$

where w_n represents the synaptic strength of the n-th synapse, by which an arrival of a single presynaptic action potential rises the synaptic conductance $g_{syn}^{(n)}$ by the amount of w_n. $\tau_{syn}^{(n)}$ is the time constant of the synaptic conductance determining how fast $g_{syn}^{(n)}$ decays as time after it rises by the impulsive effect of the single presynaptic action potential. $i(n)$ is the indexing number to represent the i-th arrival spike for the n-th synapse. $t_{i(n)}$ represents the time instant when the i-th spike arrives at the n-th synapse. (3.103) is similar to (3.95) for the membrane potential, but this is for the synaptic conductance. With these $g_{syn}^{(n)}$, the membrane potential dynamics of a general neuron model may be described as

$$C\frac{dV}{dt} = -i_{ion} + \sum_{n=1}^{N} g_{syn}^{(n)}(V - E_{syn}), \tag{3.104}$$

where i_{ion} represents the sum of ion channel currents, E_{syn} the reversal potential of the synapse. This ODE may be solved numerically together with the ODE in (3.103).

The delta function $\delta(t)$ is just a mathematical concept, and it does not suit for numerical computations. Considering a case where the time t is discretized by a step Δt, how can we represent the delta function $\delta(t - a)$ located at $t = a$? One can confirm numerically that introducing a pulse with its width Δt and the height $1/\Delta t$ during a simulation does not give you a satisfactory outcome. Because of this, ODE simulations with the delta functions, such as for (3.103), it is required to use a technique to avoid a direct use of the delta function. For example, when we solve (3.103) numerically, the following procedure is often taken:

- Use the following ODE unless the presynaptic neuron of the n-th synapse fire:

$$\frac{dg_{syn}^{(n)}}{dt} = -\frac{g_{syn}^{(n)}}{\tau_{syn}^{(n)}} \tag{3.105}$$

- Shift the synaptic conductance instantaneously by the amount of w_n as follows if a firing of the presynaptic neuron is detetcted at $t = t_{i(n)}$:

$$g_{syn}^{(n)}(t_{i(n)}) \rightarrow g_{syn}^{(n)}(t_{i(n)}) + w_n \tag{3.106}$$

or

$$g_{syn}^{(n)}(t_{i(n)} + \Delta t) \rightarrow g_{syn}^{(n)}(t_{i(n)} + \Delta t) + w_n \tag{3.107}$$

3.3.1 Reciprocal Inhibition and Alternating Rhythms

A pair of muscle groups both actuating a joint of a body-skeletal system is called *antagonistic muscle*. One is a flexor muscle group, and the other is an extensor muscle group. If a muscle force of the flexor is greater than that of the extensor, the joint is flexed. It is extended if a muscle force of the extensor is greater than that of the flexor. These two muscle groups may contract alternately during a simple periodic motion of the joint such as during locomotion. They can also contract simultaneously if it is required to make the joint stiff and rigid.

Contractions of a muscle fiber are controlled by action potentials of the corresponding (innervating) motoneurons in the spinal cord. Indeed, motoneurons are the final output devices that deliver neural commands to outer environment of the central nervous system, the muscles. The muscle contracts more strongly as the firing frequency of the motoneuron that innervating the muscle increases. Alternating contractions of the antagonistic muscles are made possible by the neural mechanism called the *reciprocal inhibition* which generates alternating cellular activities between two groups of motoneurons, one for the flexor muscle group and the other for the extensor muscle group. Action potential generations of the motoneurons are

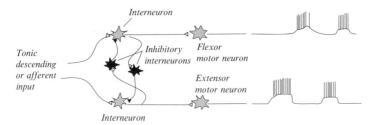

Fig. 3.12 A simple model of neuronal network with reciprocal inhibition. It is a classical model of a neuronal network that can generate rhythmic and alternating neuronal bursting activity accounting for neural basis of alternating contractions of flexor and extensor muscles

controlled by spinal interneurons as well as neurons in the brain. It has been considered that, during locomotion of animals, spinal neural networks of interneurons alone can generate autonomously rhythmic, coordinated, and alternating spiking dynamics of populations of neurons. In such a case, one population of interneurons fire and generate spikes to increase activity of the corresponding population of motoneurons, say the flexor motoneurons. During this period, the other population of interneurons for the extensor motoneurons may cease their firings. Similarly, when the extensor interneurons fire to drive the extensor motoneurons, the flexor interneurons may cease their firings. The alternating contractions of the antagonistic muscles during locomotion can be realized by repeating this process.

Figure 3.12 illustrates a model of a neuronal network that can reproduce the alternating spiking dynamics of a pair of interneuron populations. We consider the model involving four populations of interneurons. Note that, in Fig. 3.12, each population of interneurons is represented by a cartoon of a single cell. Cells in one population are excitatory presynaptic neurons of postsynaptic motoneurons that stimulate the flexor muscle fibers for the joint of the body-skeletal system. Those in the other population are also excitatory presynaptic neurons of motoneurons that stimulate the extensor muscle fibers for the same joint. In the model of Fig. 3.12, these two populations synaptically interact via the reciprocal inhibition, for which the other two populations of inhibitory interneurons are required. As in Fig. 3.12, one is the population of postsynaptic interneurons of the flexor interneurons with excitatory synapses, and these interneurons are presynaptic of the population of the extensor interneurons with inhibitory synapses. The other is the population of postsynaptic interneurons of the extensor interneurons also with excitatory synapses, and they are presynaptic of the population of the flexor interneurons with inhibitory synapses. This network possesses all essential mechanisms necessary for the reciprocal inhibition. The flexor and extensor interneurons may receive excitatory tonic (non-rhythmic) inputs from the higher brain (the brainstem), inducing firings of these interneurons. Even if this network has complete symmetry between the flexor populations and the extensor populations, asymmetry in an initial condition of every neuron may lead to the alternating dynamics between the flexor populations and the extensor populations.

IDE Modeling: Rhythm Generator Spinal Network Model: Search ModelDB by "Rybak" to Find Rybak_Rhythm_Generator_Network.isml

$$C\frac{dV_i}{dt} = -I^i_{Na} - I^i_K - I^i_L - I^i_{synE},$$

$$C\frac{dV^r_i}{dt} = -I^i_{Na} - I^i_{NaP} - I^i_K - I^i_L - I^i_{synE} - I^i_{synI},$$

$$I^i_{synE} = g^i_{synE}(V_i - E^i_{synE}),$$

$$I^i_{synI} = g^i_{synI}(V_i - E^i_{synI}),$$

$$\frac{dg^i_{synE}}{dt} = -\frac{g^i_{synE}}{\tau^i_{synE}} + \sum_{j(n)} w_{ji}\delta(t - t_{j(n)}),$$

$$\frac{dg^i_{synI}}{dt} = -\frac{g^i_{synI}}{\tau^i_{synI}} + \sum_{j(n)} (-w_{ji})\delta(t - t_{j(n)}),$$

where V_i is the membrane potential of the i-th inhibitory interneuron, V^r_i the that of the i-th excitatory interneurons. The inhibitory interneurons are regular spiking neurons, which generate spikes in response to excitatory inputs. The excitatory interneurons are rhythmic bursting neurons, and they show rhythmic burst with their alternating active and silent phases like in the pancreatic β cells that we studied previously. The rhythmic bursting neurons possess I_{NaP} channel current, referred to as persistent sodium channel. I_{NaP} includes slowly inactivating h-gate that provide

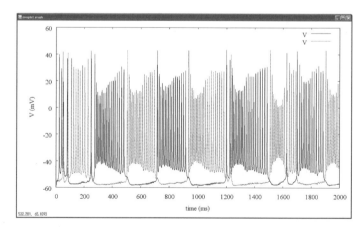

Fig. 3.13 IDE modeling of Rybak spinal cord rhythm generating neuronal network. The model includes 40 spontaneous bursting interneurons and 40 non-bursting inhibitory interneurons. *Solid* and *dotted curves* are the action potentials of interneurons from different populations, one from a flexor interneuron and the other from an extensor interneuron

a slow dynamics required for the bursting dynamics. I^i_{synE} and I^i_{synI} represent, respectively, the excitatory and inhibitory synaptic currents for the i-th neuron. g^i_{synE} and g^i_{synI} are, respectively, the excitatory and inhibitory synaptic conductances for the i-th neuron. The rhythm generator neuronal network model by Rybak et al. (2006) assumes 20 neurons for each of the four populations necessary for the reciprocal inhibitions. Two inhibitory populations include the regular spiking neurons, and two excitatory populations includes the rhythmic bursting neurons. Each of the rhythmic bursting neurons receives a sequence of tonic, non-rhythmic excitatory synaptic inputs from a higher brain, which induces the rhythmic burst. The mechanism of reciprocal inhibition allows the model to exhibit the alternating bursting dynamics as shown in Fig. 3.13. Note that it is not obvious at all if the reciprocal inhibition always guarantees the alternating bursting between the two populations. Indeed, the emergence of the alternating dynamics is dynamic, and thus, the model can also show non-alternating dynamics depending on values of parameters in the model, such as intensity of the tonic inputs, the synaptic strength, as well as parameter values that affect characteristics of ion channels.

Chapter 4
Application of ISIDE to Create Models

In this chapter, usage of the *insilico* platform is demonstrated. The *insilico* platform is composed of three blocks, i.e. *insilico*ML, *insilico*IDE and *insilico*DB. *Insilico*ML (ISML) (Asai et al. 2008) is a language specification based on XML to describe mathematical models of physiological functions. *Insilico*IDE (ISIDE) (Kawazu et al. 2007; Suzuki et al. 2008, 2009) is a software program on which users can simulate and/or create a model with graphical representations corresponding to the concept of ISML, such as modules and edges. ISIDE also has a command line interface to manipulate large scale models based on Python, which is a powerful script computer language. ISIDE exports ISML models into C++ source codes, CellML format and FreeFEM++ format for further analysis or simulation. *Insilico*Sim (ISSim) (Heien et al. 2009), which is a part of ISIDE, is a simulator for models written in ISML. *Insilico*DB is formed from three databases, i.e. database of ISML models (Model DB), timeseries data (Time-series DB) and morphological data (Morphology DB). These databases are open to the public at the website www.physiome.jp.

4.1   Brief Introduction of ISML

*insilico*ML (ISML) is a language specification based on XML to describe mathematical models of physiological functions. Details of the ISML specification are described in Chap. 5. Here we briefly illustrate the idea of ISML. In a model written in ISML, each of physiological entities is represented as a *module*. Each module is quantitatively characterized by several *physical-quantities* defined in the module. Physical-quantities are used to represent constant/variable parameters as well as dynamical variables used in the definition of time evolution of the state of the module. ISML has a capability to describe not only ordinary differential equations, but also partial differential equations, rules for multi agent systems, and to integrate morphological and timeseries data into mathematical models.

Structural and functional relationships between two modules are defined by *edges* spanned between the modules. For example, if one module physically includes another module (e.g. cell membrane includes mitochondria) they are

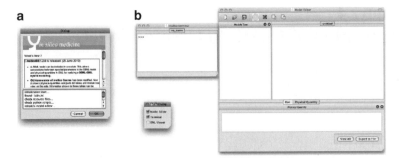

Fig. 4.1 Launching *insilico*IDE (ISIDE). (**a**) Splash screen of ISIDE. (**b**) *Upper left panel* is *insilico* terminal in which Python commands are available to edit a model, and *right* is *insilico* canvas on which users can create a model with a graphical user interface. *Lower left panel* is a launcher of the terminal, canvas and XML viewer

connected by a *structural edge* with a keyword "include". If a module quantitatively affects to another module (e.g. an ionic current flowing a channel on a cell membrane changes the membrane potential), the two modules are linked by a *functional edge* with a keyword, for example, "hyperpolarize", which allows one module who wants to refer a value of a physical-quantity defined in the other module to utilize the value in itself. There are *in-ports* and *out-ports* on a module to receive and export the numerical information. A functional edge is connected at a port on a module.

Edges are directed. A module connected to a tail of a structure edge is considered to be a sublayer of a module which is pointed by a head of the edge. Thus, in other words, there is a parent–child relationship. A functional edge is directing from an out-port to an in-port. A functional edge does not define hierarchical (parent–child) relationships, but network-type relationships.

A concept to make a kind of package of a physiological function has been introduced to ISML, which is called *capsulation*, in order to enhance the model sharing in the public domain. By the capsulation, several modules acting together as a certain physiological function are encapsulated by a capsule module. All connections to (from) an encapsulated module from (to) outside of the capsule must go through the capsule to secure the independence of the encapsulated modules. By this isolation of modules, it becomes easy to utilize the encapsulated modules in other part of the model or in other models. See Sect. 5 for more details.

4.2 Getting Started

*Insilico*IDE (ISIDE) is available on the web site www.physiome.jp, which currently supports Windows XP, Mac OSX 10.5, and Mac OSX 10.6. Once a user launches the application then a splash screen (Fig. 4.1a) appears, and then white canvas and a terminal open as shown in Fig. 4.1b. Users can immediately start to create a new model on the canvas. Users can download ISML models from the public model

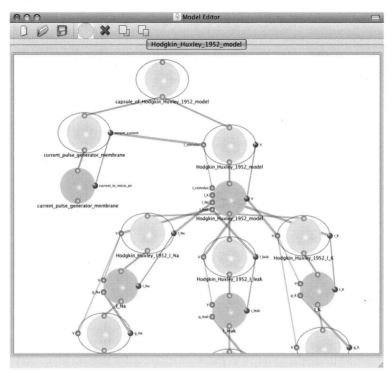

Fig. 4.2 Graphical representation of a model. Each module is represented as a *yellow* or *blue ball*. A *circle rounded ball* is a capsule module. *Lines* connecting modules represent edges. There are a couple of types for edges. See text for details

database ModelDB at www.physiome.jp. Let us download a Hodgkin–Huxley model (Hodgkin and Huxley 1952) with pulse input stimulus. Users can search the model on ModelDB with a keyword such as "hodgkin" to find the model named "Hodgkin_Huxley_1952_model.isml". Then load the model on the *insilico* canvas.

Modules in the model are represented as circles as shown in Fig. 4.2. Double clicking on a module toggles showing or hiding modules in a lower level in the hierarchical structure of the model. Edges are represented by several types of lines, e.g. gray, thick pink, etc., according to the type of edges, such as structural and functional. A user can move modules by dragging cursor with pressing left mouse button on a module, and span an edge by dragging cursor with pressing left mouse button from an out-port which is represented as a small blue ball at a right side of a module, to an in-port, a small red ball at a left side of another module.

Clicking a right mouse button on a module shows a command menu. From the list, you can open dialogs to edit the information of ports, physical-quantities and so on.

To run a simulation of the model, at first launch *insilico*Sim (ISSim) from a menu bar of ISIDE, and open the model. After setting up configurations for a simulation, such as algorithm for numerical integration, time step, sampling interval and so on,

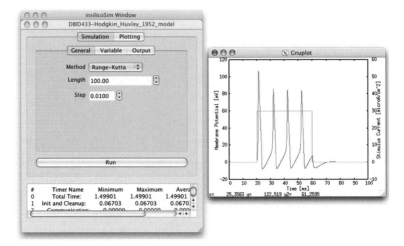

Fig. 4.3 *Left* is an *insilico*Sim control panel on which settings for a simulation and plotting can be done. *Right* is a graph of the simulation result drawn by gnuplot working behind *insilico*Sim

at the Simulation tag, press Run button. A while later, the simulation will finish, and the user can plot the simulation result from the Plotting tab on the ISSim window (Fig. 4.3). There is another way to run simulation of the ISML model. ISIDE can export the ISML model into C++ codes which includes numerical integration algorithm as well. Then as usual way for performing simulation, users just need to compile the C++ codes and execute the binary simulation program. Then text files including data time series are obtained.

4.3 Model Creations and Simulations on ISIDE

Model creation on *insilico* platform is carried out using the four components of the platform, i.e. ISIDE, ISML, ISSim and ISDB (Fig. 4.4). In this chapter, we demonstrate how to create models on ISIDE, and to run simulations. With taking Hodgkin Huxley model as an example, basic procedures to create a model is illustrated in several sections from Sect. 4.3.1. After that a couple of techniques helpful to create variety of models are introduced.

4.3.1 Hodgkin–Huxley Model

The Hodgkin–Huxley model (Hodgkin and Huxley 1952) is a well-known conductance based model of an excitation of neuron membrane. Since the details of the model is explained in Sects. 2.3 and 2.4, only a brief summary is shown here. The membrane potential can be calculated by an integration of three major currents, i.e. voltage-dependent persistent Potassium ion (K^+) current, voltage-dependent

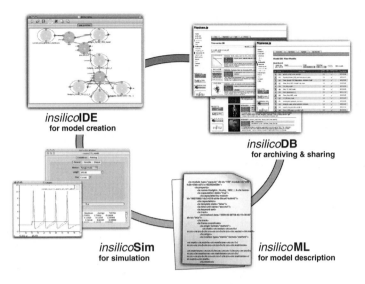

Fig. 4.4 The *insilico* platform is composed of four building blocks, i.e. *insilico*IDE: a model builder with graphical user interface, and *insilico*Sim: a simulator, *insilico*DB: databases of models, morphological data and timeseries data, and *insilico*ML: a model description language

transient Sodium ion (Na$^+$), and a leak current which is considered to be mainly carried by Chloride ion (Cl$^-$), and at the same time to represent other channels which are not described explicitly. The derivative of the membrane potential V_m is represented by a summation of aforementioned currents;

$$C\frac{dV_m}{dt} = -I_K - I_{Na} - I_{leak} + I_{ext},\tag{4.1}$$

where C, I_k, I_{Na}, I_{leak} and I_{ext} represent, respectively, membrane capacitance, K$^+$ current, Na$^+$ current, leak current, and current additionally given externally. Each ion current is described by the potential difference between the cell membrane V_m and the equilibrium potential of the ion (E_{ion}) multiplied by a conductance (g_{ion}) where $ion =$ K$^+$, Na$^+$ or *leak*.

$$I_{ion} = g_{ion}(V_m - E_{ion}).\tag{4.2}$$

The conductance for the leak current is a constant, while the conductances for K$^+$ and Na$^+$ are modeled as the time dependent variables in Hodgkin–Huxley model. By denoting the maximum conductance (all gates are open) as $\bar{g}_{ion}$, the conductance g_{ion} can be written with its activation ratio r which is called a gating variable.

$$g_{ion} = \bar{g}_{ion}r.\tag{4.3}$$

The gating variable evolves according to the following equation

$$\frac{dr}{dt} = \alpha_r(1 - r) - \beta_r r,\tag{4.4}$$

where α and β are the transition rates which are non-linear functions of membrane potential V_m. K^+ current is controlled by four identical activation gates n, similarly Na^+ current by three activation gates m and one inactivation gate h.

Hence the ion currents in the Hodgkin–Huxley model are described as follows:

$$I_K = g_K(V_m - E_K), \quad g_K = \bar{g}_K n^4,$$

$$I_{Na} = g_{Na}(V_m - E_{Na}), \quad g_{Na} = \bar{g}_{Na} m^3 h,$$

$$I_{leak} = g_{leak}(V_m - E_{leak}), \tag{4.5}$$

with the rate "constants"

$$\frac{dm}{dt} = \alpha_m(V_m)(1 - m) - \beta_m(V_m)m,$$

$$\frac{dn}{dt} = \alpha_n(V_m)(1 - n) - \beta_n(V_m)n,$$

$$\frac{dh}{dt} = \alpha_h(V_m)(1 - h) - \beta_h(V_m)h. \tag{4.6}$$

4.3.2 Modularity of a Model

Before starting model creation, it is important to think of the way of structuration of the modeling target physiological phenomena. In other words, we need to decide what kind of modules we create, and how we make structural and functional association among them. ISML allows users to make a tree structure of modules in the sense of the physical relationship (structure) of physiological functions, and a network structure in the sense of functional relationships among modules. The former is defined by structural or logical edges and the latter is by functional edges.

In a tree structure of modules, the top module is called "root module", which means there are no higher layer for the module. A model can take multiple root modules. In other words, there can exist multiple trees on the *insilico* canvas. In the case of Fig. 4.5, there are two root modules, one is a capsule module of stimulus current generator and the other is a capsule module of the Hodgkin–Huxley model.

In the case of the Hodgkin–Huxley model described above, the main actors are the membrane potential V_m, four ionic currents I_K, I_{Na}, I_L, I_{ext}, and three gating variables m, n and h. The dynamics of the membrane potential depends on the four currents, hence we can consider that the membrane potential is *composed* of these currents. The dynamics of the sodium current, for example, depends on the two gating variable m and h. These considerations define the structure of the model which looks like Fig. 4.6a.

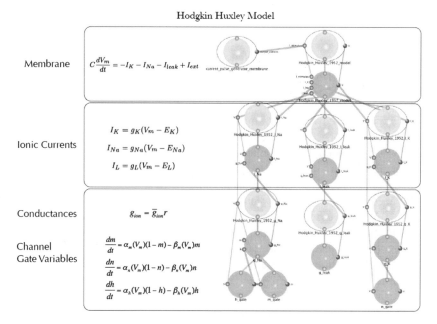

Fig. 4.5 An example of a hierarchical and modular structure of a Hodgkin–Huxley model

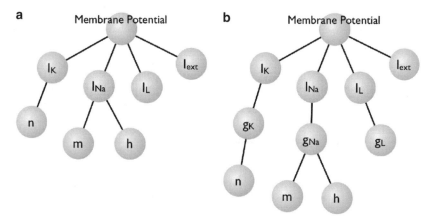

Fig. 4.6 There are various ways to structuralize a model. These examples show possible physical relationships among modules in a Hodgkin–Huxley model

The way to structuralize the model is not unique. It is natural that the model structure depends on the intentions and view points of model creators. For example, the Hodgkin–Huxley model is a rather simple model, but still we can find another way to structuralize as shown in Fig. 4.6b, in which a conductance is considered as an entity of each ionic current, and the conductance is a function of the gating variables. Such diversity of the structure of one model can be rather important scientifically, when we compare models. In the following sections, we create a Hodgkin–Huxley model as shown in Fig. 4.5.

4.3.3 Conductance Modules

Let us begin with creating conductance module g_K. First of all, let us create a module named g_K in which the conductance $g_K = \bar{g}_K n^4$ is defined. By clicking an icon on the menu bar or selecting a menu "Make Module..." in a context menu appearing by right mouse button click, a dialog to create a module pops up (Fig. 4.7). There are two types of module to create. One is *container*, and the other is *functional-unit*. The former defines a conceptual module which bundles several modules as one group without defining any physical-quantities in it. The latter acts as a functional entity which can be quantitatively characterized by several physical quantities. In this case the g_K module must be the functional-unit type.

We need to define three physical-quantities in the module, i.e. $\bar{g}_K$, g_K and n (Fig. 4.8). When a user wants to add a physical-quantity in a module, select "Edit Physical Quantity..." from a context menu appearing by right clicking on the target module. Then a dialog shown in Fig. 4.8 emerges on which users can edit physical-quantities. There are six types of physical-quantities such as *state*, *variable-parameter*, *static-parameter*, *func-expression*, *nominal* and *morphology*.

For example, $\bar{g}_K$ is defined as a static-parameter type physical-quantity among six types, which represents a time invariant parameter through the simulation. The concrete implementation of the physical-quantity can be given by several ways

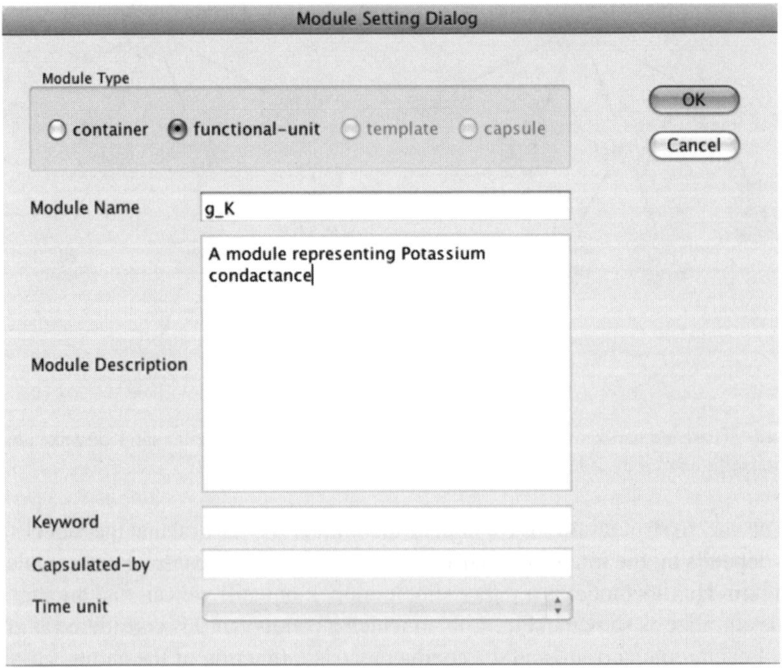

Fig. 4.7 A dialog to create a module. When a user creates a new module, it is required to select the type of the module from container or functional-unit, and to give the name of the module, description keywords and so on

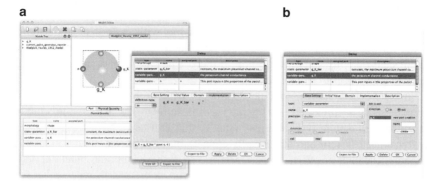

Fig. 4.8 A module representing the conductance for the potassium channel current and a dialog to edit physical-quantities. The module has one in-port named n and one out-port g_K. On the dialog, several basic properties of the physical-quantity such as name, type, unit, description, etc. as well as its mathematical expression and port assignment can be edited

including describing with mathematical expressions. Here users can select one of 11 types to give the implementation, i.e. *ode* (ordinary differential equation), *pde* (partial differential equation), *dde* (delay differential equation), *sde* (stochastic differential equation), *de* (difference equation), *ae* (algebraic equation), *func-expression, assign, graph, conditional* and *loop*. $\bar{g}_K$ is implemented as an algebraic equation (ae) defining $\bar{g}_K = 36 \, \mathrm{mS}^{-1}/\mathrm{cm}^2$.

As we saw in creation of the physical-quantity $\bar{g}_K$, to define one physical-quantity users need to select at first the type of physical-quantity, and then the type of implementation, and finally give a concrete expression, i.e. implementation, such as an equation.

Next let us define a physical-quantity g_K in the physical-quantity edit dialog, which is a *variable-parameter* type physical-quantity, meaning a parameter whose value can change during simulation, because g_K is a function of n which is a dynamical variable (thus a function of time). The implementation is given as an algebraic equation (ae) $g_K = \bar{g}_K n^4$ (Fig. 4.8a). The output of this module is the value of the conductance g_K which is used in the calculation of the potassium channel current. Hence we also need to create an out-port to export the value of g_K, and to make an association between the out-port and the physical-quantity g_K. In the dialog to edit physical-quantities, it is also possible to make an association between the physical-quantity to an out-port if free out-ports, i.e. out-ports which have no association to a physical-quantity yet, exist. If there is no free out-ports, it is also possible to create a new out-port and make association (Fig. 4.8b). The creation of an out-port and making association to a physical-quantity can be done also on a dialog to edit port properties.

In the equation for g_K, the gating variable n is used. According to the idea of the model structure (Fig. 4.5), we will create another module to define the dynamics of n later. At this moment let us define n as a variable-parameter type physical-quantity, which receives the value from the other module through an in-port. The implementation type of the physical-quantity is "assign" and sub-type is given as "port". We also need to create an in-port to receive the value of n, and to make an association between the in-port and the physical-quantity n so that the physical-quantity can refer the value arrived to the in-port. In a similar way with creation of an out-port as we did for the physical-quantity g_K, it is possible to make an association between the physical-quantity and a free in-port. If there is no free in-port, it is also possible to create a new in-port and make association.

Now we create another module n_gate, in which the dynamics of the gating variable $dn/dt = \alpha_n(V_m)(1 - n) - \beta_n(V_m)n$ is defined. The actors in this module are n which is a *state* type physical-quantity that is used to represent the state of the dynamical system and that usually appears in a differential equation, and α_n, β_n and V_m which are variable-parameters. V_m is the membrane potential and is considered as an input to this module. An in-port is created to receive V_m from other module, and the implementation of the physical-quantity V_m is "assign" type to define the association to the in-port. α_n and β_n are defined as variable-parameter type and implemented as algebraic-equations $\alpha_n = 0.01(10 - V_m)/(\exp((10 - V_m)/10) - 1)$ and $\beta_n = 0.125\exp(-V_m/80)$. n is a state type physical quantity and is implemented as an ordinary differential equation $dn/dt = \alpha_n(1 - n) - \beta_n n$. The output of the module is the value of n. So we create an out-port associated to n.

We consider that the module n_gate is located at hierarchically sublayer of the module g_K. We link these two modules by a structural edge from the top of the n_gate to the bottom of the g_K as shown in Fig. 4.9 left. The structural edge is directed from the n_gate module to the g_K module. Notice that the position (top or bottom) on a module where the structural edge attaches is a crucial key to represent a parent–child relationship between two modules, i.e., the n_gate module is a child of the g_K module.

The in-port of the module g_K is supposed to receive the value of the physical-quantity n from the module n_gate. So let us span an functional edge between the out-port of the n_gate and the in-port of the g_K, indicating the flow of the information. The functional edge is also directed from the out-port to the in-port, but does not define parent–child relationships. There is no input to the in-port of the module n_gate to receive the value of the membrane potential, however for the moment let us leave it as an open in-port, and later we will back to this port.

Finally, let us encapsulate these two modules. Encapsulation is a process to make a kind of package of modules for better reusability. In this case, two modules g_K and n_gate are working together to calculate the potassium conductance. We can consider these two modules as a package which outputs the value of conductance calculated for a given value of the membrane potential that will be given to this module from an input to the package. The concept is shown in Fig. 4.10. This is rather simple example, however if we think of a physiological function modeled by tens of modules, each time we reuse these modules, we need to link edges to each module which is inconvenient (Fig. 4.10a). However once we bundle these modules as a package (encapsulation), all input and output interfaces are summarized on the capsule (Fig. 4.10b). Then we can utilize the encapsulated modules as one functional unit regardless of the details of the inside (Fig. 4.10c), which can enhance the reusability of this set of modules. This is the idea of "encapsulation" on the *insilico* platform.

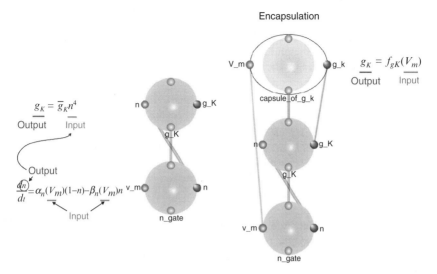

Fig. 4.9 (*Left*) A model of the potassium channel conductance. Two modules are linked by a structural edge. They also have a functional-edge from the out-port of the child module to the in-port of the parent module. The value n calculated in **n_gate** module is used in the module **g_K**. (*Right*) Encapsulated potassium channel conductance model. The capsule module has one in-port to receive the information of the membrane potential V_m, and one out-port to export the value of the conductance g_K, as an interface of the encapsulated modules. The ports on the capsule module and ports on the module inside of the capsule module are connected by forwarding edges. By encapsulation, the module **capsule_of_g_K** can be used as a g_K calculator. A user just needs to know the meanings of the input (V_m) and output (g_K), and not the detailed implementation of the function $f_{gK}(\cdot)$, when the user uses this module as one of building blocks of a model

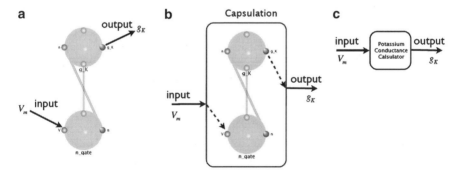

Fig. 4.10 The idea of the capsulation on *insilico* platform. (**a**) There are modules working together as a physiological function, which have input and output. (**b**) Capsulation is a process to make a kind of package of modules. By the capsulation, all inputs and outputs can be summarized on the capsule module. (**c**) The capsule module can be used as one functional unit. Users do not need to know the details of inside of the capsule

When a user applies the command *encapsulation* in a right button context menu at a module, **g_K** in this case, a capsule module is automatically created at top of the module, which is linked to the module **g_K** with a capsular edge. The representation is not like Fig. 4.10b because of a practical reason. In this example, as input/output interfaces to this encapsulated modules, one in-port to receive the value of the membrane potential V_m and one out-port to export the value of g_K are necessary. Once a user creates these ports on the capsule module, link each port to the corresponding port, such as the in-port on the module n_gate and the out-port on the module g_K, by a forwarding edge (Fig. 4.9 right). A forwarding edge defines information transfer between a capsule module and modules encapsulated by the capsule, hence links between an in-port (out-port) on the capsule module and an in-port (out-port) on modules encapsulated by the capsule module. On this point, a forwarding edge is not like a functional edge which links from an out-port to an in-port of modules. By the encapsulation, these two modules are packed and can be easily treated as a model representing the potassium conductance.

As summarized in Fig. 4.9, the structure of the potassium conductance model we created above can also be understood in the view point of mathematical equations. The left hand side of the equation $g_K = \bar{g}_K n^4$ is considered as the output of the module g_K. In the right hand side, there are two variables. We decided that $\bar{g}\_K$ is defined in the same module as a static-parameter, and the dynamics of n is defined in the other module n_gate, and in this module we created only the variable-parameter to receive the value from n_gate. Of course it is possible to define n as a state type physical-quantity with ordinary differential equation in the module g_K instead of creating the other module. In that case, the hierarchical structure becomes different from the one that we created. The process of encapsulation is corresponding to the abstract mathematical functions such as $g_K = f_{gK}(V_m)$ to calculate the potassium conductance as a function of V_m. The precise definition of the function $f_{gK}(\cdot)$ is implemented in the modules under the capsule module.

Exercise 4.1. Create sodium and leak conductance modules:

Create sodium and leak conductance modules in a similar way with the potassium conductance modules. The conductance g_{leak} is considered as a time invariant value. So only one static-parameter type physical-quantity is defined in the module. Use the following parameter values and formulae. The leak conductance is $0.3 \, \mathrm{mS}^{-1}/\mathrm{cm}^2$. The maximum sodium conductance $\bar{g}_{Na} = 120 \, \mathrm{mS}^{-1}/\mathrm{cm}^2$. The sodium conductance depends on two gating variables m and h. Let us make a module for each of them. Transition rates for each of m and n gate are modeled as follows: α_m and β_m for the m gate are defined as variable-parameter type and implemented as $\alpha_m = 0.1(25 - V_m)/(\exp((25 - V_m)/10) - 1)$ and $\beta_n = 4\exp(-V_m/18)$. For the h gate, $\alpha_h = 0.07(-V_m)/\exp(20)$ and $\beta_h = 1/(\exp((-V_m + 30)/10) + 1)$.

4.3.4 Current Modules

A module representing the potassium current includes the following physical-quantities: current I_K which is the output of this module, membrane potential V_m and conductance g_K as variable-type physical-quantities which are inputs, and the equilibrium potential E_K as a static-parameter. Since V_m and g_K are calculated in other modules, we create two in-ports to receive these values and make association between ports and physical-quantities. That is, the implementation of these physical-quantities are "assign" type which define the reference to the in-ports. E_K is implemented with an algebraic equation $E_K = -12\,\text{mV}$. The output of this module is the potassium current I_K, so the module has one out-port associated to I_K, which defines a mathematical implementation $I_K = g_K(V_m - E_K)$.

This module uses the capsule module of g_K as a building block. We put the capsule_g_K module as a child module of I_K by linking with a structural edge. Then span a functional edge from the out-port of the capsule module to the in-port of I_K to define the information flow of the potassium conductance. Finally, we again encapsulate the module I_K and create an in-port to receive the information of V_m and an out-port to output the value of I_K as the interface on the new capsule module. Then link the in-port V_m on the new capsule module to the in-port of the module I_K with a forwarding edge. At the same time, do not forget to link from the in-port of the capsule module to the in-port of the capsule module of g_K with a forwarding edge to give the value of V_m. Similarly, span a forwarding edge from the out-port on the module I_K to the out-port on the new capsule module (Fig. 4.11).

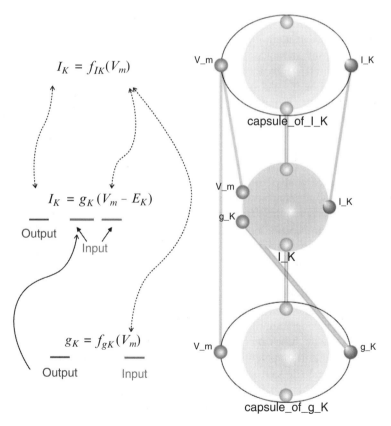

Fig. 4.11 The encapsulated model of the potassium current I_K which uses the output (value of g_K) of the model of the potassium channel conductance g_K

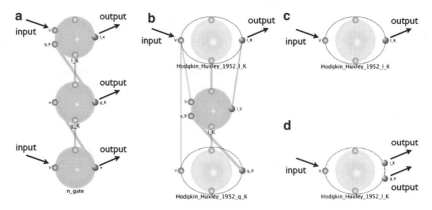

Fig. 4.12 Encapsulation can be nested. In other words, a capsule module can include several other capsule modules as its structure member. This rule makes a modeling simpler. (**a**) A potassium current model without using encapsulation, and (**b**) with encapsulation. (**c**) Simplification of the visual representation of the model. (**d**) To make the output g_K accessible from outside of the capsule, we need to add one more out-port on the capsule module and to link those out-ports by a forwarding-edge

Notice that the capsulation can be nested. In other words, a capsule module can include other capsule modules as its structure member. This is one beneficial point of the encapsulation. If we do not use the encapsulation, the potassium current model looks like Fig. 4.12a. If we utilize this model in other place of other model, we need to give the value of V_m as inputs to two modules. All out-ports such as I_K, g_K and n are accessible. When we introduce encapsulation as shown in Fig. 4.12b, all we need to do to calculate the value of I_K is to give the input value V_m to the top capsule module. And only the out-port I_K of the capsule module is available. The way to utilize this model becomes much simpler than the case without encapsulation. Fig. 4.12b is conceptually equivalent to Fig. 4.12c with simple visualization. Due to this simplification, now we cannot access to other values such as g_K. In a case that such a value is required from outside of the capsule module, we just add an out-port on the top capsule module, and link a forwarding edge from the out-port related to the value g_K to the new out-port on the capsule module. Then g_K becomes accessible (Fig. 4.12d).

Exercise 4.2. Create sodium and leak current modules:

Create sodium and leak current modules in a similar way with the potassium current modules using sodium and leak conductance modules created in the previous exercise. Use the following parameter values and formulae. For sodium current modules, the equilibrium potential $E_{Na} = 115$ mV (static-parameter with algebraic equation (ae)), and the current is defined as $I_{Na} = g_{Na}(V_m - E_{Na})$ (variable-parameter with ae). For leak current, $E_{leak} = 10.613$ mV (static-parameter with ae) and $I_{leak} = g_{leak}(V_m - E_{leak})$.

4.3.5 Membrane Module

The membrane module calculates the dynamics of the membrane potential V_m which is modeled as a state type physical-quantity with the mathematical implementation $dV_m/dt = -I_K - I_{Na} - I_{leak} + I_{ext}$. The module membrane_potential is located at top of the three ionic current models created in the previous section, and linked with them by structural edges. We create variable-parameter type physical-quantities I_K, I_{Na}, I_{leak} and I_{ext} in this module, which are associated to respective in-ports to receive the values calculated in the other modules. Let us link the functional edges from the out-ports of the capsule modules of ionic currents I_K, I_{Na} and I_{leak} to the corresponding in-ports on the module of the membrane potential. The output of this module is the membrane potential V_m, so the corresponding out-port is needed to be created and associated to the physical-quantity V_m. The value of V_m is used in the ionic current modules to calculate the currents and conductances. We also need to span functional edges from the out-port of this module to the in-port of the capsule modules representing ionic currents. I_{ext} represents the external stimulus current given to the membrane, which is the input signal to the Hodgkin–Huxley model.

Let us encapsulate the module representing the membrane. Creating an in-port for receiving I_{ext} and an out-port to exporting the value of the membrane potential V_m on the capsule module, and linking forwarding edges to these ports, we finally built up the Hodgkin–Huxley model as shown in Fig. 4.13.

4.3.6 Add an External Stimulus Current Generator Model

This Hodgkin–Huxley model requires an external current injecting to the membrane. Since there is an in-port on the capsule module at the top of the model, a

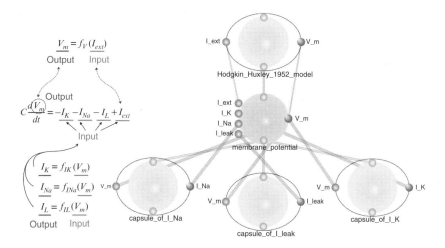

Fig. 4.13 The encapsulated model of the membrane potential which uses the models of the sodium, potassium and leak currents

functional edge must be connected to the in-port otherwise the value of the physical-quantity associated to the in-port cannot be defined for running the simulation. It is of course possible to create a new module of a stimulus current generator, however, since there are already several stimulus current generators in the model database (ModelDB), let us reuse one of them. Here we will use a step current generator as an example. At the website www.physiome.jp/modeldb, users can find a step current generator model by searching with a keyword "step" for model name. Once downloading the ISML model from the database, it is possible to add the new model to the currently edited model on the *insilico* canvas. By this procedure we can reuse the existing models as building blocks of new models.

When an existing model is added to the *insilico* canvas, the top module of the model appears on the canvas beside the currently edited modules. In our example, a module current_step_generator_membrane appears as shown in Fig. 4.14. The added module has an out-port from which a current is exported. We can span a functional edge from the out-port to the in-port of the capsule module of the Hodgkin–Huxley model to complete a simulation executable model.

The step current generator model includes one module under the top capsule module, which is numerically characterized by static parameter type physical-quantities representing initial current, step current height, step onset time, and a variable parameter type physical quantity representing the output stimulus current. The output stimulus current is defined using conditional branching and algebraic equations as follows: if *time > step onset time* then *output current = initial value + step height*, else *output current = initial value*.

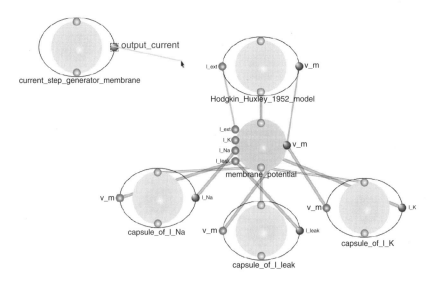

Fig. 4.14 Instead of creating a new external current generator module, users can add an existing model to *insilico* canvas beside the currently edited model to reuse other model as one of building blocks. All users need to do is to link functional and structural edges between newly added modules and the currently edited modules

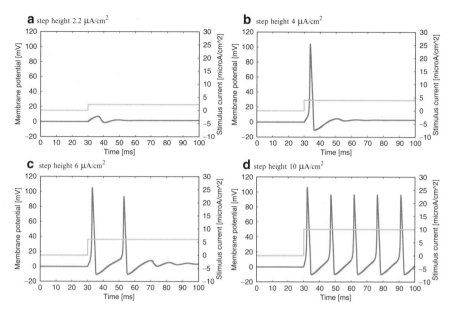

Fig. 4.15 Examples of simulation results of the Hodgkin–Huxley model for various values of the stimulus intensity

The example of simulation results of the Hodgkin–Huxley model with various values of stimulus current step height are shown in Fig. 4.15. The results shown in Fig. 4.15 were obtained by ISSim of ISIDE. For running the simulation of the model created on ISIDE, once we need to save the model into a file in ISML format. Then ISSim loads the model file. Once the model is loaded users need to set several configuration for running simulation, such as algorithm and time step of numerical integration, simulation duration and so on. If the model includes many physical-quantities, it is possible to select physical-quantities that you want to save the simulation result as timeseries data. After setting up all configuration, press Run button, then the simulation starts. Plotting tab provides the interface to plot the simulation result. Users can select physical-quantities to be used for ordinate and abscissa of a graph. Range and label of the plot can be set. Pressing Plot button, a window with graph pops up.

To investigate the dependency of the dynamics of the membrane potential of the Hodgkin–Huxley model to the stimulus current intensity, i.e. the height of the step current, as shown in Fig. 4.15, we need to follow the steps below. (1) Change the value of the physical-quantity on the dialog for editing physical-quantity of ISIDE. (2) Apply the change. (3) Save the model. (4) Re-run the simulation on ISSim. (5) Re-plot the graph. Users do not need to reload the model on ISSim after changing the values of parameters in the model on ISIDE. Each time Run button is pressed, ISSim check the model and parse the contents of the model for a new simulation. ISSim also remember the configurations for simulation and plotting.

4 Application of ISIDE to Create Models

4.3.7 Voltage Clamp Experiment

The voltage clamp is a common technique used in cellular-electrophysiological experiments to control the transmembrane voltage to test the electrogenic activity of membrane proteins. Hence it can be also applied to analyze the voltage-gated ionic currents of the neuron membrane. The Hodgkin–Huxley model is a suitable model to simulate the application of the voltage clamp to the membrane. As we created ionic current modules in Sect. 4.3.4, an ionic current module calculates its current based on the membrane potential as well as the conductance. If we replace the voltage input from the membrane module to the current module by a step voltage, we can simulate the voltage clamp experiments.

First of all, let us find a step voltage generator model in the database (ModelDB). We can find the model **voltage_step_generator** by searching with a keyword "step" in model name. Figure 4.16a shows the Hodgkin–Huxley model and a voltage step generator module added to the *insilico* canvas at left bottom corner. There are three ionic current modules, i.e. sodium, potassium and leak currents. Each of them has an in-port to receive a voltage. To make a voltage clamp model, firstly we remove other modules than capsule modules representing the ionic currents from the Hodgkin–Huxley model. To leave the ionic current modules on the canvas and remove other modules, we need to detach the ionic current modules from the tree structure. Select a structural edge between the membrane module and the capsule module of the potassium current, and delete it. Similarly other structural edges linked to ionic current modules must be removed. By this operation, three capsule module modeling the ionic currents becomes independent from the original tree structure of modules. Then at the top capsule module of the Hodgkin–Huxley model, select a command "Remove all descendant modules" from the right button context menu. All modules under the top capsule module are removed. Similarly stimulus current generator modules can be removed, since we do not use them in the voltage clamp simulation.

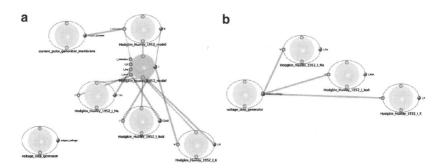

Fig. 4.16 Models for the voltage clamp simulation experiment. (**a**) The original Hodgkin–Huxley model at *right upper part* and a step voltage generator module at *left bottom*. (**b**) A model of the voltage clamp experiment. The voltage generator module gives a step voltage to the three ionic current modules

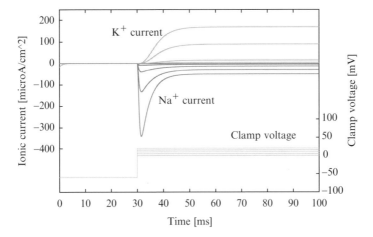

Fig. 4.17 Ionic currents and step voltages observed during the simulation of the voltage clamp experiment. The simulations show the fast dynamics of the sodium current and the slow dynamics of the potassium current. The clamp voltage changes from -60 mV to 0, 5, 10, 15, 20 mV. The larger the step voltage is, the larger the absolute value of sodium and potassium currents' variations are

Next let us link functional edges from the out-port of the step voltage generator module to the in-port of each of the ionic current modules (Fig. 4.16b). By this modification, each ionic current module receives the step voltage instead of the membrane potential.

Figure 4.17 shows the results of the voltage clamp simulations for various values of step voltages. It can be observed that at the onset of the step voltage from -60 mV, there is rapid dynamics of the sodium current representing opening and closing of the gates, and slow movement of the potassium current started. The leak current is not shown in the figure but it flows in proportion to the given voltage.

4.3.8 Replacement of Ionic Current Modules

Let us modify the Hodgkin–Huxley model by installing an additional sodium current module which is used in another model. This example is illustrated here to demonstrate how to receive the full benefit of encapsulation and model database in the process of creation of a new model.

At first let us download a model of a neuron in charge of rhythm generation in spinal neural circuitry involved in locomotor pattern generation. Searching with a keyword "rybak" for model name, we can find a model named Rybak_RG_neuron_V.isml. Originally a computational model of the mammalian spinal cord circuitry incorporating a two-level central pattern generator (CPG) with separate half-centre rhythm generator (RG) and pattern formation (PF) networks has been proposed by Rybak et al. (2006). The above ISML model is the RG part of the CPG model.

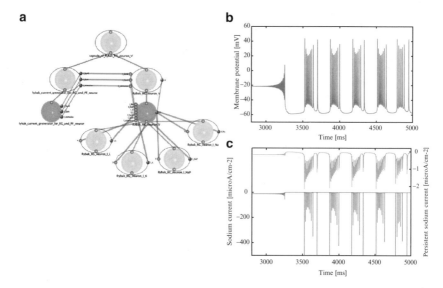

Fig. 4.18 Rhythm generator (RG) neuron model which is a component constituting a locomotor central pattern generator circuitry model proposed by Rybak et al. (2006). (**a**) Representation on *insilico* canvas of ISIDE. (**b**) The membrane potential dynamics calculated by ISSim. (**c**) The sodium and persistent sodium currents plotted on left and right ordinates, respectively

This model looks like as shown in Fig. 4.18a on ISIDE. Figure 4.18b shows an example simulation result of the membrane potential of the RG neuron performed by ISSim. The dynamics of the membrane potential of the RG neuron is described by the following ODE:

$$\frac{dV_m}{dt} = \frac{1}{c}\left\{-I_K - I_{Na} - I_{NaP} - I_{leak} - I_{synE} - I_{synI} + I_{ext}\right\}, \qquad (4.7)$$

where I_K, I_{Na}, I_{leak} are the potassium, sodium and leak currents, respectively, similar to those in the Hodgkin–Huxley model. I_{synE} and I_{synI} are excitatory and inhibitory synaptic inputs, respectively. In this example, since the single RG neuron is isolated from other neurons, these synaptic inputs are set to zero. I_{ext} is an external stimulus current, set to $0.3\,\mu\text{A/cm}^2$. I_{NaP} is the persistent sodium current which plays an important role for the endogenous rhythmogenic properties of RG neurons. As shown in Fig. 4.18c, the sodium current and persistent sodium current show very different behavior to each other during the simulation. The sodium current immediately returns back to near zero after each spike, however the persistent sodium current does not.

We can investigate the electrogenic property of the channels by the voltage clamp method. Figure 4.19a shows the voltage clamp model using the sodium and persistent sodium current modules in Rybak_RG_neuron_V model and a step voltage generator. As we can see in the simulation result shown in Fig. 4.19b, persistent sodium current exhibits much slower response than that of the sodium current.

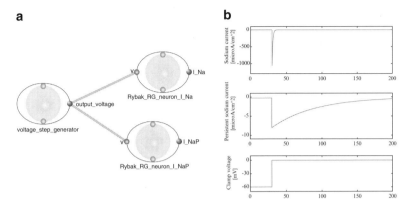

Fig. 4.19 Voltage clamp of sodium and persistent sodium channels on the RG neuron. (**a**) The voltage clamp model made on ISIDE. Two ionic current capsule modules, i.e. sodium and persistent sodium currents, are extracted from Rybak_RG_neuron_V model and step voltage generator was added. (**b**) A result of voltage clamp simulation experiment

Now let us modify the Hodgkin–Huxley model using the persistent sodium current in the Rybak RG neuron model. Namely, we will add the persistent sodium current model to the membrane of the Hodgkin–Huxley model, and investigate the changes in the dynamics of the membrane potential. Note that, this is a hypothetical in silico experiment, which has no physiological background. Nevertheless, it would be worthwhile to perform this for better understanding of how the encapsulation works.

First of all let us leave only the capsule module of the persistent sodium current in the Rybak RG neuron model and remove all other modules on ISIDE. And add the Hodgkin–Huxley model. Then we need to modify the module representing the membrane to slot the persistent sodium current modules into the Hodgkin–Huxley model. Let us create an in-port to receive the persistent sodium current on the membrane module, and a variable-parameter type physical-quantity associated to the in-port to receive the value. Let us call the physical-quantity I_NaP. Then we modify the ODE for the membrane potential of the Hodgkin–Huxley model as follows:

$$\frac{dV_m}{dt} = \frac{1}{C}\left\{-I_K - I_{Na} - I_{NaP} - I_{leak} + I_{ext}\right\}. \qquad (4.8)$$

We add I_{NaP} in the right hand side of the ODE. Let us link the capsule module of the persistent sodium current to the membrane module with a structural edge, and link the out-port of the persistent sodium current and the new in-port of the membrane module, that we created right now.

We need to be careful a little bit on the value of the membrane potential. To calculate the persistent sodium current, the membrane potential is required. In the Hodgkin–Huxley model, the resting potential is set to zero for convenience, but it is about $-60\,\mathrm{mV}$ in the Rybak RG neuron. We need to adapt the membrane potential to each other. Let us create a kind of adaptor of the membrane potential as a module, which calculates $V_{RG} = V_{HH} - 60$, where V_{RG} is used in the persistent sodium

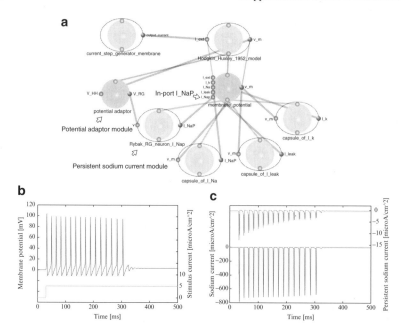

Fig. 4.20 The modified Hodgkin–Huxley model. (**a**) Representation of model structure on ISIDE. (**b**) The dynamics of the membrane potential. (**c**) The sodium and persistent sodium currents

modules originated in the Rybak RG neuron, and V_{HH} is the membrane potential used in the Hodgkin–Huxley model. Hence the adapter module has an in-port to receive V_{HH} and an out-port to export V_{RG}. Finally we link a functional edge from the out-port of the membrane module to the in-port of the adaptor module, and from the out-port of the adapter module to the in-port of the persistent sodium current module.

The representation of the modified Hodgkin–Huxley model structure on ISIDE is shown in Fig. 4.20a. Modules and an in-port added in the above modification are indicated by arrows. The dynamics of its membrane potential are shown in Fig. 4.20b. The intensity of the step stimulus current is $5 \, \mu A/cm^2$. The spike generations ceases at some point after the onset of the step stimulus current. Figure 4.20c shows the dynamics of the sodium and persistent sodium currents. The amplitude of the persistent sodium current diminishes as time passing after the onset of the step current, while that of the sodium current almost does not change.

This modification of the model and investigation of the behavior of the model are just a kind of fun as mentioned. But this example shows the principal idea to create your own models based on the existing models in the model database.

4.4 Neural Network Model

In this section, how to create neural network models on ISIDE by reusing existing models is illustrated. When a neural network model is rather simple, it is possible to create such a model on ISIDE in a graphical way with clicks and drags of mouse operations as described so far. However a neural network model can be very complicated. For example, a model can include thousands of neurons and synapses. Then to create large scale models only with graphical user interface is almost impossible. We need to think of another way to create large scale models.

ISIDE provides a command line interface to build a model. The interface looks like a terminal on which Python script can be executed. It means that script syntaxes such as for-loop and while-loop are available in a model creation process. This is very helpful to automate repetitive operations and systematic model constructions. Using the mixture methods between graphical and command line interfaces, the creation of large scale models becomes practically possible.

4.4.1 Two-Coupled BVP Model

Let us begin with a simple neural network which is composed of two coupled Bonhoeffer–van der Pol (BVP) models (Asai et al. 2003). We consider a simple hard-wired model of the central pattern generator for locomotion consisting of two identical oscillators connected by reciprocal inhibition. In the model, each oscillator acting as a neural half-center controlling movement of a single limb, either left or right, is modeled by the BVP equations which receives a control input corresponding to a flow of descending signals in spinal cord from higher motor centers.

The single BVP equations (FitzHugh 1961) can be considered as a two-dimensional simplification of the Hodgkin–Huxley model of spike generation, and is described as follows:

$$\frac{dv}{dt} = c\left(v - \frac{v^3}{3} - w + I_{DC}\right),$$

$$\frac{dw}{dt} = \frac{1}{c}(v - bw + a), \tag{4.9}$$

where v and w represent, respectively, the membrane potential and refractoriness of the neuron. Equations (4.9) are mathematically equivalent to the FHN model (FitzHugh 1961; Nagumo et al. 1962) considered in Sect. 2.5, with different expression of the formulae.

We can find an ISML model of the single BVP in the model database, whose name is single_BVP_with_step_current_stimulus.isml. Figure 4.21a shows the graphical structure of the model on ISIDE. There are two modules representing the membrane potential and refractoriness beneath the capsule module BVP. The first and second ODE of (4.9) is defined, respectively, in the module of the membrane potential and refractoriness. The two modules at leftward of Fig. 4.21a represent the step current generator. According to the intensity of the external current I_{DC}, the

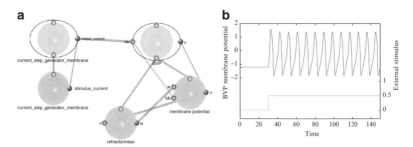

Fig. 4.21 A single BVP model. (**a**) Graphical representation of the model on ISIDE. (**b**) An example of a simulation result. The parameters in (4.9) are set as follows: $a = 0.7, b = 0.675$, $c = 1.75, I_{DC} = 0.5$

membrane can act either as an excitable or oscillatory medium. An example of a simulation result is shown in Fig. 4.21b.

Using the above BVP model, let us make a two coupled BVP model as a model of the hardwired half-center model of the central pattern generator for locomotion. The model can be written mathematically as follows:

$$\frac{dv_1}{dt} = c\left(v_1 - \frac{v_1^3}{3} - w_1 + I_{DC1}\right) + \delta(v_2 - v_1),$$

$$\frac{dw_1}{dt} = \frac{1}{c}(v_1 - bw_1 + a) + \epsilon v_2,$$

$$\frac{dv_2}{dt} = c\left(v_2 - \frac{v_2^3}{3} - w_2 + I_{DC2}\right) + \delta(v_1 - v_2),$$

$$\frac{dw_2}{dt} = \frac{1}{c}(v_2 - bw_2 + a) + \epsilon v_1. \tag{4.10}$$

Each of the variable sets (v_1, w_1) and (v_2, w_2) corresponds to the state of a single BVP model. There are two coupling pathways between two single BVP models. The terms $\delta(v_2 - v_1)$ and $\delta(v_1 - v_2)$ are considered as an electrical coupling between two BVP models, and ϵv_2 and ϵv_1 are symmetric contralateral projections from the excitatory element v_i $(i = 1, 2)$ on one side to the refractoriness (inhibitory) element w_j $(j = 2, 1)$ on the opposite side. Let us implement this model on ISIDE using the existing single BVP. We need to modify the single BVP model to be able to receive external signals for creating couplings between two single BVP modules.

First of all, let us modify the membrane potential module in which the dynamics of v is defined as an ODE. What we need to do is to add an in-port to receive the external current I_ext, to make a new physical-quantity I_ext to assign the value coming from the in-port, and to modify the ODE of v. We can already carry out these manipulations by GUI interface on ISIDE. The same manipulations can also be done with command-line interface. For further application of the processing on the command-line, let us try to use the command-line interface here.

On the *insilico* terminal users can use Python script. Python-supported *insilico* platform APIs are provided on the *insilico* platform to access to model entities.

There are several handlers such as module_handler, edge_handler, port_handler, physicalquantity_handler and so on. The handlers have many methods to deal with an ISML document of a model. For example, an edge_handler has methods to manage the edge related information, and similarly port_handler has methods to manage the port related information. Thus, even there are methods with the same name such as **create**, edge_handler creates an edge, and port_hander creates a port.

To create an in-port, we execute the following Python API using the port_handler on the terminal.

```
>>> port_handler.create('582c0fc2-b2ca-4aeb-b550-3ca43efa5ea4', 'in', 'Iext')
```

The first argument is the ID of the module on which the new port is created. The module ID can be obtained as follows. Select "copy" menu in the context menu popped up by pressing a right mouse button on the target module (which is **membrane potential** in this case). Or select the module and simply press Ctrl (or Command) + C. Then users can get the module ID in a clipboard buffer. After this, the module ID appears as a string if users paste it on the *insilico* terminal. Note that if users paste it on the *insilico* canvas, the module specified by the module ID is duplicated and the copy of the module appears on the canvas as a new module with a new ID. The second argument is the direction of the port, which takes either "in" or "out". The last argument is the name of the port. The command returns a value (5 in this case) which is the ID of the port. The modifications done by commands on the terminal is immediately reflected to the model and graphical representation on the canvas (Fig. 4.22).

Next we create a physical-quantity to be associated to the in-port. Using physicalquantity_handler, we create a physical-quantity without implementation at first. The type of the physical-quantity is "variable-parameter".

```
>>> physicalquantity_handler.create_with_type('582c0fc2-b2ca-4aeb-b550-
    3ca43efa5ea4', 'variable-parameter', 'Iext')
```

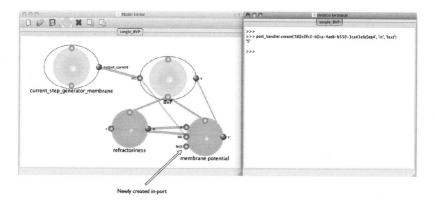

Newly created in-port

Fig. 4.22 Command-line interface to create an in-port on a module. Using port_handler we can manipulate attributes of ports. In this example, it creates a new in-port. The first argument is the ID of the module, the second is the direction of the port, "in" or "out" and the last is the name of the port

This command returns 8 which is the ID of the newly created physical-quantity. The first argument of the command is the module ID in which the physical-quantity is created. The second argument is the type of the physical-quantity, and the last is the name of the physical-quantity.

Let us give a concrete definition to the physical-quantity.

```
>>> definition_handler.set_implementation_definition('582c0fc2-b2ca-4aeb-b550-
    3ca43efa5ea4', '8', '0', '5', 'assign', 'port')
```

This command means to "assign" the "port" whose ID=5 to the physical-quantity with ID=8 in the module with ID=582c0fc2-b2ca-4aeb-b550-3ca43efa5ea4. The third argument is set to 0 indicating a substructural position in the implementation. Substructures of an implementation is tightly related to the ISML specification and are introduced for describing complicated definition, including conditional branchings (IF – ELSE IF – ELSE) and loop-structure (FOR, WHILE), of the physical-quantity. See Sect. 5.4.2 for more details. When we do not need to concern about such substructures in an implementation, we can simply set zero for the third argument in the command.

To modify the ODE defined on v can be done as follows. Firstly we need take the physical-quantity ID of v.

```
>>> physicalquantity_handler.get_id_with_name('582c0fc2-b2ca-4aeb-b550-
    3ca43efa5ea4','v')
```

The first argument of the command is the module ID in which we look for the physical-quantity and the second is the name of the target physical-quantity. This returns 2 in this case, which is the physical-quantity ID. We will add the term I_{ext} in the right hand side of the current ODE defined for v as follows

$$\frac{dv}{dt} = c\left(v - \frac{v^3}{3} - w + I_{DC}\right) + I_{ext}. \tag{4.11}$$

Since the definition of the equation must given by MathML format, we need to convert this to MathML.

```
>>> cstr2mml.set_ns_prefix('m')
>>> mml = cstr2mml.cstr2mml('diff(v, time)=c*(v-pow(v,3)/3-w+I_DC)+I_ext')
```

cstr2mml is a package included in *insilico* platform APIs to convert from C language like expression of equations to MathML format. The first command set_ns_prefix is to set the namespace prefix for MathML part for better discriminability of MathML part from ISML part. In this case, all MathML related tags have m: namespace prefix. The conversion result is as follows:

```
<m:math>
<m:apply>
  <m:eq/>
  <m:apply>
    <m:diff/>
    <m:bvar>
      <m:ci>time</m:ci>
    </m:bvar>
```

```
      <m:ci>v</m:ci>
    </m:apply>
  <m:apply>
    <m:plus/>
    <m:apply>
      <m:times/>
      <m:ci>c</m:ci>
      <m:apply>
        <m:plus/>
        <m:apply>
          <m:minus/>
          <m:apply>
            <m:minus/>
            <m:ci>v</m:ci>
            <m:apply>
              <m:divide/>
              <m:apply>
                <m:power/>
                <m:ci>v</m:ci>
                <m:cn>3</m:cn>
              </m:apply>
              <m:cn>3</m:cn>
            </m:apply>
          </m:apply>
          <m:ci>w</m:ci>
        </m:apply>
        <m:ci>I_DC</m:ci>
      </m:apply>
      <m:ci>I_ext</m:ci>
    </m:apply>
  </m:apply>
</m:apply>
</m:math>
```

Now we set and define this equation as a new implementation of the physical-quantity v,

```
>>> definition_handler.set_implementation_definition('582c0fc2-b2ca-4aeb-b550-
    3ca43efa5ea4', '2', '0', mml, 'ode', '')
```

This is the same function as we used above to define "assign" type implementation, but with different argument values. The MathML expression is given at the fourth argument instead of port ID. The fifth argument specifies type "ode" and the sixth that specifies additional information of the type, is empty. For example, in a case of type "pde" instead of "ode", a category of the PDE must be selected from "elliptic", "parabolic", "hyperbolic" and "others", and given to the sixth argument as the additional information. But since there is no such additional feature for ODE to be defined, the sixth argument must be empty.

We can verify the result of modifications done so far on ISIDE on the physical quantity edit dialog (Fig. 4.23).

Next we need to create an in-port on the capsule module BVP to receive an external current to be forwarded to the in-port Iext of the membrane potential module, and link them with a forwarding-edge.

```
>>> port_handler.create('458aa0ca-f041-43b5-91bf-582655b4c747', 'in', 'Iext_v')
'5'
>>> builder_handler.link_ports('458aa0ca-f041-43b5-91bf-582655b4c747', '5',
    '582c0fc2-b2ca-4aeb-b550-3ca43efa5ea4', '5', 'forwarding')
'043c8e27-73c6-4952-938d-065e16201825'
```

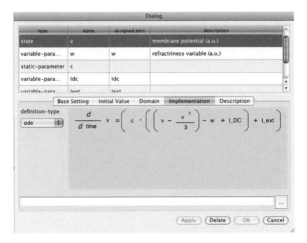

Fig. 4.23 The physical-quantity created by Python API commands on the terminal can be seen in the dialog for physical-quantity edit. The implementation of the ODE for v is re-defined with a term I_{ext}, which is the physical-quantity added by a terminal command with assignment to the new in-port

The first command creates an in-port on the capsule module with its name "Iext_v", giving back '5' as a return value. The second command spans a forwarding edge from the in-port (port ID=5) of the capsule module (module ID=458aa0ca-f041-43b5-91bf-582655b4c747) to the in-port (port ID=5) of the membrane potential module (module ID=582c0fc2-b2ca-4aeb-b550-3ca43efa5ea4). This returns an ID of the edge created by this command. In this example, we used the builder_handler to create a new edge instead of edge_handler, because builder_handler has many of compound functions for convenience. That is, if we use edge_handler instead of builder_handler, at first we need to create a new edge, set the type of the edge which is "forwarding" and define the head and tail of the edge with their module ID and port ID. These four commands are bundled in one command "link_ports" of the builder_handler.

We finish all modifications on the membrane potential module. Now we come to the modification of the refractoriness module. The procedures is very similar to what we did above, i.e. to add an in-port to receive the external current I_ext and make a new physical-quantity I_ext to assign the value coming from the in-port, and then modify the ODE of w. Then we make a new in-port on the capsule module, and a forwarding edge linking the in-port of the capsule module and the in-port of the refractoriness module. The ODE of w will be

$$\frac{dw}{dt} = \frac{1}{c}(v - bw + a) + I_{ext}. \tag{4.12}$$

The commands to modify the refractoriness module are shown below in which we use the ID of the refractoriness module "2288c88f-4571-4fd7-b8df-77a4002ca542".

```
>>> port_handler.create('2288c88f-4571-4fd7-b8df-77a4002ca542', 'in', 'Iext')
'3'
```

```
>>> physicalquantity_handler.create_with_type('2288c88f-4571-4fd7-b8df-
    77a4002ca542', 'variable-parameter', 'Iext')
'7'
>>> definition_handler.set_implementation_definition('2288c88f-4571-4fd7-b8df-
    77a4002ca542', '7', '0', '3', 'assign', 'port')
True
>>> mml = cstr2mml.cstr2mml('diff(w, time)=(v-b*w+a)/c+I_ext')
>>> definition_handler.set_implementation_definition('2288c88f-4571-4fd7-b8df-
    77a4002ca542', '2', '0', mml, 'ode', '')
True
```

The followings are for modifying the capsule module and linking a forwarding
edge between the refractoriness module and the capsule module.

```
>>> port_handler.create('458aa0ca-f041-43b5-91bf-582655b4c747', 'in', 'Iext_w')
'6'
>>> builder_handler.link_ports('458aa0ca-f041-43b5-91bf-582655b4c747', '6',
    '2288c88f-4571-4fd7-b8df-77a4002ca542', '3', 'forwarding')
'ae24b8aa-6e8e-494e-ad26-a8802a43d27e'
```

Finally we obtained the single BVP model modified so that it can receive external
currents to the membrane potential module and the refractoriness module as shown
in Fig. 4.24.

Let us create coupling modules. In the two coupled BVP model there are two
types of the couplings. One is the electrical coupling between membranes, which
is so called a gap connection (see Sect. 3.1), and contralateral projections from the
membrane potential module to the refractoriness module on the opposite side.

A current flowing through the electrical coupling between two BVP models is
calculated by the difference of membrane potential of two neurons multiplied by a
conductance of the coupling path. The current flowing from the second neuron to
the first neuron is calculated by

$$current_1 = \delta(v_2 - v_1), \tag{4.13}$$

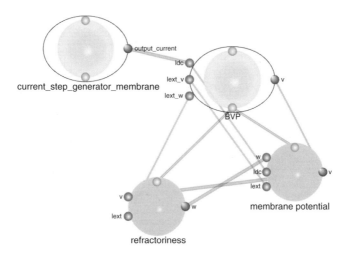

Fig. 4.24 The single BVP model modified to receive external currents to the membrane potential
module and refractoriness module

where δ is a conductance of the gap junction. Similarly the current the second neuron receives through the gap connection is

$$current_2 = \delta(v_1 - v_2).\qquad(4.14)$$

Since the module representing the electrical coupling needs to calculate the difference between membrane potentials of two neurons interacting by the gap junction, it is required to have two in-ports to receive membrane potentials from two neurons. The module also should have two out-ports to export the current flowing through the gap junction. The current exported from the out-port named Current 1 in the right bottom module in Fig. 4.25 has the same absolute value with the one from the out-port Current 2 but with an opposite sign.

The contralateral projection from the membrane module of the first neuron to the refractoriness module of the second neuron is calculated by

$$current_1 = \epsilon v_2,\qquad(4.15)$$

where ϵ corresponds to a synaptic efficacy of the projection. Similarly the current that the second neuron receives via the projection is

$$current_2 = \epsilon v_1.\qquad(4.16)$$

The module structure is similar to the module for the electrical coupling. Finally we encapsulate these two modules. On the terminal the following command

```
>>> builder_handler.encapsulate(['e13195bc-4f1c-445f-8049-0dd3f06e7fcd',
    '01619b6f-98a7-43d3-928c-f2cc9702dc41'])
```

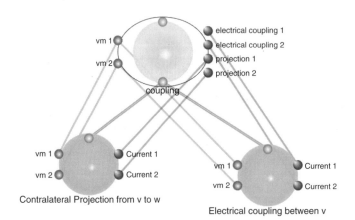

Fig. 4.25 The modules representing couplings between two BVPs. We consider an electrical coupling between membrane and contralateral projections from the membrane module to the refractoriness module on the other side

makes a capsule module grouping these two modules used for coupling together. The argument is a list of module IDs that we want to encapsulate in the same capsule module. The command returns the ID of the created capsule module. After making in- and out-ports and forwarding edges, we obtained the coupling model as shown in Fig. 4.25.

We need two BVP models for making coupling. We can simply duplicate the modified single BVP that we prepared above. On *insilico* canvas, select the copy from the right button context menu or Ctrl (or Command) + C on a target module. Then click anywhere on the white space where you want to put the copy, and select paste from the context menu or Ctrl (or Command) + V. Then the module you selected and all modules in sublayers are copied there. Instead of clicking white place, if you click (select) a module and then perform a paste, the duplicated modules will belong to the selected module as a substructure of the selected module. Of course this can also be done with a command line. On the terminal, execute the following command

```
>>> builder_handler.copy_subtree('71fe3eff-a55e-4a3d-a51a-983fc5e9dc4d','root')
```

The first argument is the target module ID to be copied, and the second is the location where we create the copy. All modules under the module with the given ID are copied with maintaining the hierarchical structure defined by edges. If a module ID is given to the second argument, the duplicated modules will belong as children to the module specified by the second argument. In the example here, 'root' is given as the second argument, which means that the copied module is pasted on the canvas as an independent tree, i.e. the top module of the copied modules will be a root module.

Let us link functional edges among modules as shown in Fig. 4.26a. Finally the two coupled BVP model is completed. Giving the model to ISSim, we can perform simulations on the model. Figure 4.26b shows two examples of the simulation results with different I_{DC} values. The model structure is bilaterally symmetric, however the symmetry can be broken in the dynamics of the membrane potentials.

Exercise 4.3. Run simulations to find peculiar behavior of the membrane potentials: In Fig. 4.26b two examples of simulation results are shown. The dynamics is asymmetric derived from the difference in the values of I_{DC1} and I_{DC2}. Examine if the symmetry of the dynamics of the membrane potentials can be broken spontaneously under the condition $I_{DC1} = I_{DC2}$.

The reader can perform the simulation in this Exercise using a downloaded ISML model with its name simple_hard_wired_central_pattern_generator_model.isml for a confirmation.

4.4.2 Three-Coupled FHN Model

As the second example, we will think of creation of the electrically coupled three FitzHugh–Nagumo (FHN) model (FitzHugh 1961; Nagumo et al. 1962) which is considered in Sect. 3.1. FHN is another name of BVP, and they are mathematically

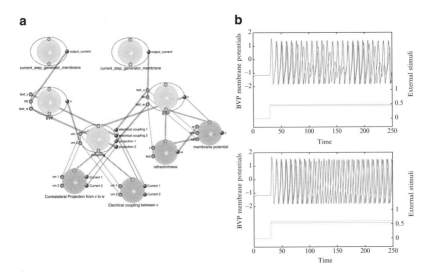

Fig. 4.26 A two-coupled BVP model. (**a**) Graphical representation of the model on ISIDE. (**b**) Examples of simulation results for different intensities of the external stimuli. The parameters in (4.10) are set as follows: $a = 0.7, b = 0.675, c = 1.75, I_{DC1} = 0.45, I_{DC2} = 0.477$ for *upper graph* and $I_{DC1} = 0.54, I_{DC2} = 0.6$ for *lower graph*

equivalent with different formulae. As illustrated in Sect. 3.1, each FHN model in the network is formulated as follows:

$$\frac{dv_n}{dt} = -v_n (v_n - a)(v_n - b) - w_n + I_n^{ext} \qquad (4.17)$$

$$\frac{dw_n}{dt} = \epsilon (v_n - cw_n)$$

where v_n and w_n represent the membrane potential and refractoriness of the membrane of the n-th neuron. I_n^{ext} is an input current applied to the n-th neuron, which is composed of the currents passing through gap junctions and externally injected current as stimulus if any.

Figure 4.27a shows the modular structure of a single FHN model with periodic stimulus current on ISIDE. The model **single-FHN_model.isml** can be found in the ModelDB or in a sample directory in the *insilico* platform package. The model includes one module representing the membrane potential in which ODE for v is defined, and the other module representing the refractoriness of the membrane in which ODE for w is defined. Only the membrane module has an in-port to receive an external current associated to a variable-parameter type physical-quantity I_n^{ext}. There is one in-port on the capsule module that receives an external current as an input to this model which is forwarded to the membrane module. This structure is principally equivalent to the one in Fig. 4.21. An example of simulation result is shown in Fig. 4.27b with a periodic pulse stimulus.

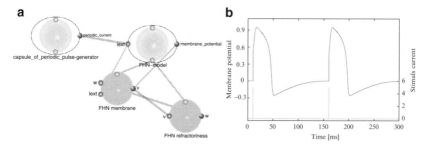

Fig. 4.27 A single FHN model. (**a**) The modular structure of a single FHN model with a stimulus current generator. The FHN model receives an external current as an input, and outputs the membrane potential. (**b**) An example of dynamics of the membrane potential responding to the periodic pulse stimulus. The pulse intensity is 5

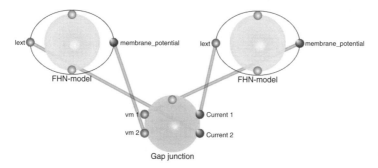

Fig. 4.28 A module of a gap junction, which has two in-ports to receive the membrane potentials of two abutting neurons and has two out-ports to return the currents through the gap junction to both of neurons. The absolute values of the Current 1 and Current 2 are the same, but with different sign

The coupling between two FHNs is modeled by the electrical coupling, i.e. gap connection. As formulated in (4.13) and (4.14), the current flowing through the gap junction is calculated by the difference of membrane potential of two adjacent neurons multiplied by a conductance of the coupling. The current flowing from the n-th neuron to m-th neuron is calculated by

$$I_{n \to m}^{gap} = G(v_n - v_m), \tag{4.18}$$

where G is a conductance of the gap junction and it is set to 0.2 here. Notice that $I_{n \to m}^{gap} = -I_{m \to n}^{gap}$.

A gap junction will be implemented as shown in Fig. 4.28. As we saw in the previous section, we need to calculate two currents $I_{n \to m}^{gap}$ and $I_{m \to n}^{gap}$.

We assume that there is an external stimulus current given only to the first neuron, and no such stimulus on the other neurons. The external current I_n^{ext} ($n = 1, 2, 3$) in (4.18) will be as follows:

$$I_1^{ext} = I_{stim} + I_{2 \to 1}^{gap} = I_{stim} + G(v_2 - v_1), \tag{4.19}$$

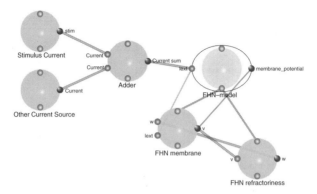

Fig. 4.29 FHN model with a current adder module which receives two currents and output a current as a result of summation of those two inputs

$$I_2^{ext} = I_{1 \to 2}^{gap} + I_{3 \to 2}^{gap} = G(v_1 - v_2) + G(v_3 - v_2), \qquad (4.20)$$

$$I_3^{ext} = I_{2 \to 3}^{gap} = G(v_2 - v_3). \qquad (4.21)$$

Since a FHN model has only one in-port to receive an external current, and the first and second neurons need to receive two currents as (4.19) and (4.21), we prepare a current adder as a module (Fig. 4.29). The adder module has two in-ports to receive currents from two sources, hence it has two variable-parameter type physical-quantities assigned to each in-port. A variable-parameter type physical-quantity I_{sum} is defined as an algebraic equation $I_{sum} = Current_1 + Current_2$, which is associated to an out-port.

Using these four building blocks, i.e. the single FHN model, the gap junction, the current adder and the stimulus current generator, a three-coupled FHN network model can be built as illustrated in Fig. 4.30a. There are modules of three FHN neurons, two gap junctions, two current adders and one stimulus current generator. Figure 4.30b shows an example of simulation result when a step current is applied to the first neuron in the network model.

On ISIDE, a user can easily duplicate modules by copy and paste operations. For example, to multiply a FHN model, a user selects the capsule module of the FHN model, and copies it. Then all modules at sublayers of the selected module are copied. By pasting it on the canvas, all copied modules are added to the canvas with preserving the structural and functional relationship among modules. Once a user has all building blocks, it is straightforward to create a network model by copy and paste and linking functional edges among them.

Since the model taken as an example above is a small network model, it is still manipulatable by hands using GUI. However if a user wants to create, for example, a network model composed of 1,000 neurons, it is impossible to carry out the model creation with manual operations. Automatic or semi-automatic ways are inevitable to create large scale models. As introduced in Sect. 4.4.1, Python commands are available on the *insilico* platform. Python is one of the most prevailing script languages for programming. To create large scale models, Python

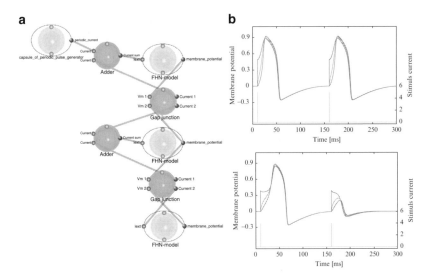

Fig. 4.30 Three-coupled FHN network model. (**a**) There are modules of three FHN neurons, two gap junctions, two current adders and one stimulus current generator. (**b**) Examples of simulation results. The periodic pulse stimulus is given to the first neuron in the network model. The pulse intensity is 5 on the *upper panel* and 3.95 on the *lower*

```
 1 #=== Settings for standalone executable script
 2 import sys, os
 3 iside_path = "/.../insilicoPlateform"
 4 sys.path.append(iside_path)
 5 os.chdir(iside_path)
 6 from initialize import *
 7 mlog_set_level(logging.INFO)
 8
 9 #=== Create a model object and model handlers
10 model = Model(uuid4_str(), isdb)
11 hh  = model.get_header_handler()
12 mh  = model.get_module_handler()
13 eh  = model.get_edge_handler()
14 cih = model.get_creator_info_handler()
15 bl  = model.get_builder()
16
```

Fig. 4.31 A Python script to create N-coupled FHN model

syntax such as "FOR" loops, conditional branching by "IF" statements and so on, are helpful for users to automate the model construction with Python-supported *insilico* platform APIs.

Figure 4.31 shows a whole script to create N-coupled FHN model using the single FHN model shown in Fig. 4.27, gap junction module (Fig. 4.28) and adder module (Fig. 4.29). The commands in API in the script are briefly explained line by line below.

```
17 #=== Number of neurons (N > 1)
18 N = 3
19
20 #=== Load building blocks
21 model.load_new_isml_model('sample/single-FHN_model.isml')
22 model.add_isml_model('sample/gap_junction.isml')
23 model.add_isml_model('sample/adder.isml')
24
25 #=== Set a model name
26 hh.set_name('%s coupled FHN model' % N)
27 cih.set_person_name('1','Yoshiyuki','','Asai')
28
29 #=== Set module IDs of the building blocks
30 [mod_fhn]  = mh.get_ids_with_name('FHN-model')
31 [mod_gap]  = mh.get_ids_with_name('Gap junction')
32 [mod_add]  = mh.get_ids_with_name('Adder')
33 [mod_stim] = mh.get_ids_with_name('capsule_of_periodic_pulse_generator')
34
35 logging.info("FHN module ID = " +mod_fhn)
36 logging.info("Gap junction module ID = " +mod_gap)
37 logging.info("Adder module ID = " +mod_add)
38 logging.info("Stimulater module ID = " +mod_stim)
39
40 #=== At first, delete edge between stimulus generator to the FHN model
41 [eid] = eh.get_ids_connected_with_tail(mod_stim, '1')
42 eh.remove(eid)
43
44 #=== Create copies of building blocks
45 fhn_ids = [mod_fhn]
46 gap_ids = [mod_gap]
47 add_ids = [mod_add]
48
49 for i in range(1,N-1):
50     logging.info('Copy turn = %s' % i)
51     fhn_ids.append(bl.copy_subtree(mod_fhn, 'root')[0])
52     gap_ids.append(bl.copy_subtree(mod_gap, 'root')[0])
53     add_ids.append(bl.copy_subtree(mod_add, 'root')[0])
54
55 #=== One more neuron model as the last neuron
56 fhn_ids.append(bl.copy_subtree(mod_fhn, 'root')[0])
57
58 #=== Link edges
59 k = 0
60 bl.link_ports(mod_stim,    '1', add_ids[k], '1', 'functional')
61 bl.link_ports(gap_ids[k], '3', add_ids[k], '2', 'functional')
62 bl.link_ports(add_ids[k], '3', fhn_ids[k], '1', 'functional')
63 bl.link_ports(fhn_ids[k], '2', gap_ids[k], '1', 'functional')
64
65 for k in range(1,N-1):
66     logging.info('Linking turn = ' + str(k))
67     bl.link_ports(gap_ids[k-1], '4', add_ids[k],   '1', 'functional')
68     bl.link_ports(gap_ids[k],   '3', add_ids[k],   '2', 'functional')
69     bl.link_ports(add_ids[k],   '3', fhn_ids[k],   '1', 'functional')
70     bl.link_ports(fhn_ids[k],   '2', gap_ids[k],   '1', 'functional')
71     bl.link_ports(fhn_ids[k],   '2', gap_ids[k-1], '2', 'functional')
72
73 k = N-1.
74 logging.info('Linking turn = ' + str(k))
75 bl.link_ports(gap_ids[k-1], '4', fhn_ids[k], '1', 'functional')
76 bl.link_ports(fhn_ids[k], '2', gap_ids[k-1], '2', 'functional')
77
78 #=== Export the model
79 model_xml = format_xml(model.dump_document())
80 save_file('%s_coupled_FHN_model.isml' % N, model_xml)
```

Fig. 4.31 (continued)

Once a user writes down all Python commands in a file, the script file can be called as a kind of plug-in of *insilico* platform from the *insilico* terminal. Click a "Open script…" in a "script" menu of the *insilico* terminal, then select a Python script in a file chooser dialog. When a user calls a script in this way, a model object, which is a fundamental object including whole ISML expression of the model, and all handlers such as module_handler and edge_handler are already prepared by ISIDE, they are available in the script.

There is another way to call the script as like a standalone executable program independently of ISIDE. For example, this script can be launched on a system's terminal by executing with Python interpreter as

```
$ python make_N_coupled_fhn.py
```

To execute the script in this way, lines 1–10 in Fig. 4.31 are necessary. The sys and os packages of Python are loaded at line 2 and path for *insilico*IDE package are set at lines 3–5. At line 6, *insilico* API is loaded. Line 7 defines a log level to show information of model creation process on the terminal. There are three log levels, i.e. DEBUG, INFO and ERROR. At line 10, a model object named "model" is created. At this moment, the model object includes only a framework of ISML without any definitions of modules, edges or other entities. Note that if users want to call this script from the *insilico* terminal, lines 1–10 should be removed or commented out.

Lines 11–15 prepare various handlers to use them for editing each part of the model. Line 18 sets a number of neurons included in this network model. At lines 21–23, each building block such as a single FHN model is loaded to the existing model. model.load_new_isml_model() cleans at first the contents of the model object, and loads a model into the model object. add_isml_model() adds a new model besides the existing model in the model object. After finishing to load, the model object includes a single FHN, one gap junction and one adder modules without linking by edges to each other. At lines 26 and 27, model properties such as name of the model and creator information are set.

To manipulate modules in the model, we need to know their module ID at first, since all modules, ports and edges are distinguished by their ID, and all handling are done based on those IDs. At lines 30–33, a module handler method to get module IDs from their name are called.

As shown in Fig. 4.27 the stimulus generator and FHN model are linked by an edge. But in the network model, the stimulus generator does not link to the FHN module directory, we need to delete the edge once. At first we need to know the edge ID by calling edge_handler method at line 41, and remove it using its edge ID at line 42. get_ids_connected_with_tail(mod_stim, '1') returns the ID of the edge that connected to a out-port with ID = 1 on a module with ID = mod_stim, which is the stimulus current module, by the tail of the edge. Since the stimulus current module has only one port, whose ID must be 1. Of course we can check the port ID by a function such as port_handler.get_ids_with_name(mod_stim, 'stim') which returns the ID of the port having the name 'stim'.

To build a N-coupled FHN model, we need to prepare N FHN neurons, $(N-1)$ gap junctions, $(N-1)$ current adder modules and one stimulus generator. We prepare a list of variables to store IDs of those objects. Lines 45–47 set the original building blocks loaded from files as the first entry. At lines 49–53, using for-loop, the building blocks are duplicated ($N - 2$) times. At line 51, a FHN model is duplicated by a command fhn_ids.append(bl.copy_subtree(mod_fhn, 'root')[0]). Since this command copies all modules under the module with ID = mod_fhn, the command returns a list of all new module IDs of duplicated modules. Here we want to take only a module ID of the top module of the tree, we take the first element from the returned list. Similarly at lines 52 and 53, the gap junction and the adder modules are duplicated. Notice that ($N - 2$) times iteration is enough here because we need ($N - 1$) gap junction and current adder modules and we already have one set when we load these building blocks at lines 21–23. Then one FHN neuron is still missing, so we need to add one more FHN neuron model as the last neuron in the serially connected network model at line 56.

Finally edges are linked among the modules. For the first FHN model, since it need to receive the current from the stimulus generator and gap junction, we need to deal with them in particular. The current adder receives currents from the stimulus generator (line 60) and the gap junction (line 61), whose output goes to the FHN model (line 62). The membrane potential goes to the gap junction as one of inputs (line 63). The rest of the parts except the last FHN model, can be treated in the same manner, hence for-loop can be applied again as written in lines 65–71. The k-th current adder ($k > 1$) receives currents from the ($k - 1$)-st and k-th gap junctions (lines 67 and 68), and output of the k-th gap junction goes to the k-th FHN model (line 69), whose membrane potential is used in the ($k - 1$)-st and k gap junctions (lines 70 and 71). For the last FHN model, mutual connections are spanned between the ($k - 1$)-st gap junction and the FHN model (lines 75 and 76).

Once users write the code listed in Fig. 4.31 into a file, users can execute it by calling from the *insilico* terminal. Be sure that the first 10 lines are commented out since we use the *insilico* terminal here. Click a "Open script..." in a "script" menu of the *insilico* terminal, then select the Python script in a file chooser dialog. It creates 3_coupled_FHN_model.isml on the current directory when $N = 3$.

Once such script has been written, it is easy to increase the number of neurons in the network model, and is possible to expand the model to very large scale. For example, when $N = 10$, the neural network looks like Fig. 4.32.

Exercise 4.4. A script to create a ten-coupled BVP network model: In Sect. 4.4.1, we considered two ways of coupling i.e. an electrical coupling and symmetric contralateral projections from the membrane module on one side to the refractoriness module on the opposite side. Modify the script in Fig. 4.31 to use the two coupling manners instead of simple electrical coupling as we saw in Sect. 4.4.2.

4.4.3 Middle Scale Neuronal Network Model

In this section, we will reproduce a hippocampal interneuronal network model proposed by Wang and Buzsáki (1996). Fast neuronal oscillations such as 20–80 Hz

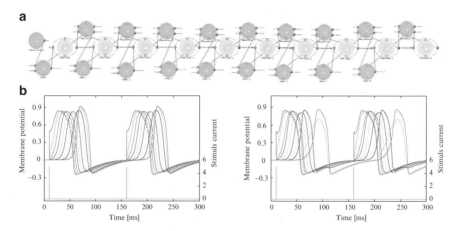

Fig. 4.32 Ten-coupled FHN network model which is created by executing a Python script using *insilico* APIs. (**a**) There are modules of ten FHN neurons, nine gap junctions, nine current adders and one stimulus current generator. (**b**) Examples of simulation results. *Left panel* shows the membrane dynamics with the uniform coupling conductances ($G = 0.2$ for all gap junctions), while *right panel* with inhomogeneous couplings

(gamma band) have been observed in the neocortex and hippocampus during behavioral arousal. They investigated the role of the GABAergic fast-spiking interneurons playing a role in the generation of gamma oscillatory activity in a randomly connected neuronal network.

In the literature, the hippocampal interneuronal network model is composed of more than 100 neurons which can be created in a similar way with the coupled FHN model shown in the previous section. Namely at first a user creates a single interneuron model, and duplicates and connects neuron models by using Python scripts. In the previous examples, we started from downloading a single neuron model and use it with some modification to create network model. This time let us create a whole single interneuron only by Python script, then expand it to the network model, although the model is also available in ModelDB with the name Wang_Buzsaki_1996_hippocampal_interneuron_model.isml.

A single interneuron model includes ten modules (eight functional-unit and two capsule modules). There are twenty four edges (thirteen functional, six structural, two capsular and three forwarding edges) as shown in Fig. 4.33a. In total the model includes forty four physical-quantities (three state, twenty three variable-parameter, ten static-parameter physical-quantities). The dynamics of the membrane potential under the existence of a direct current with intensity $1\,\mu A$ is shown in Fig. 4.33b as an example.

The entire Python script to create a single interneuron is shown below. See the comments inserted in the program for the explanation of each step. The first 16 lines are for standalone execution of the script. In a case that users call this script from the *insilico* terminal, the first 16 lines should be commented out or removed.

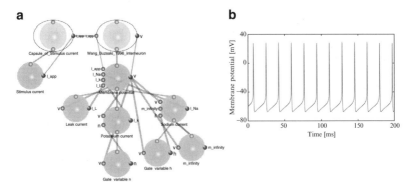

Fig. 4.33 A single interneuron model. (**a**) The modular structure of a single interneuron model with a stimulus current generator. (**b**) An example of simulation result with a direct current 1 μA given to the neuron model. The time course of the membrane potential is shown.

```
 1 # ------------------------------------------------------------------ Initialize
 2 # Settings for standalone executable Python script
 3 import sys,os
 4
 5 # Absolute path to the insilico package
 6 iside_path="/.../insilicoIDE-1.x"
 7 # Current directory is used for specifying the file name to save the model
 8 current_dir=os.getcwd() + "/"
 9 sys.path.append(iside_path)
10 os.chdir(iside_path)
11
12 from initialize import *
13 mlog_set_level(logging.INFO)
14 cstr2mml.set_ns_prefix("m")
15
16 model = Model(uuid4_str(), isdb)
17
18 # ------------------------------------------------- Create model handlers
19 hh = model.get_header_handler()
20 mh = model.get_module_handler()
21 ph = model.get_port_handler()
22 eh = model.get_edge_handler()
23 pqh = model.get_physicalquantity_handler()
24 dh = model.get_definition_implementation_handler()
25 dih = model.get_definition_initialvalue_handler()
26 hh = model.get_header_handler()
27 aih = model.get_article_info_handler()
28 cih = model.get_creator_info_handler()
29 uh = model.get_unit_handler()
30 bl = model.get_builder()
31
32 # --------------------------------------------- Create user defined units
33 unit_list = {} # "unit-code:unit-id" dictionary
34
35 ## milli-second
36 code = "ms"
37 unit_list[code] = uh.create("milli-second")
38 uh.set_element(unit_list[code], "second", "", "milli")
39
40 ## per_milli-second
41 code = "/ms"
42 unit_list[code] = uh.create("per_milli-second")
43 uh.set_element(unit_list[code], unit_list["ms"], "-1", "")
```

```
44
45 ## volt
46 code ="V"
47 unit_list[code] = uh.create("volt")
48 uh.set_element(unit_list[code], "kilogram", "", "")
49 uh.set_element(unit_list[code], "metre", "2", "")
50 uh.set_element(unit_list[code], "second", "-3", "")
51 uh.set_element(unit_list[code], "ampere", "-1", "")
52
53 ## milli-volt
54 code = "mV"
55 unit_list[code] = uh.create("milli-volt")
56 uh.set_element(unit_list[code], unit_list["V"], "0", "milli")
57
58 ## simens
59 code = "S"
60 unit_list[code] = uh.create("simens")
61 uh.set_element(unit_list[code], "ampere", "", "")
62 uh.set_element(unit_list[code], unit_list["V"], "-1", "")
63
64 ## milli_siemens_per_centi_meter2
65 code = "mS/cm2"
66 unit_list[code] = uh.create("milli_siemens_per_centi_meter2")
67 uh.set_element(unit_list[code], unit_list["S"], "", "milli")
68 uh.set_element(unit_list[code], "metere", "-2", "centi")
69
70 ## micro_ampere_per_centi_meter2
71 code = "uA/cm2"
72 unit_list[code] = uh.create("micro_ampere_per_centi_meter2")
73 uh.set_element(unit_list[code], "ampere", "", "micro")
74 uh.set_element(unit_list[code], "metere", "-2", "centi")
75
76 # -------------------------------------------------------- Header settings
77 # Single neuron module name
78 single_name = "Wang_Buzsaki_1996_interneuron"
79
80 # Set the model name
81 hh.set_name("Wang_Buzsaki_1996_interneuron_model")
82
83 # Set description
84 hh.set_description("Hippocampal interneuron model module")
85
86 # Set article information.
87 # Using a pubmed ID, the article information can be retrieved from the
        website of PubMed.
88 pubmed_id = "8815919"
89 aid = aih.create()
90 aih.setup_pubmedinfo_from_togows(aid, pubmed_id)
91
92 # Settings for numerical integration
93 # As numerical integration algorithm either Euler method (euler) or
94 # 4th-order Runge--Kutta method (4th-rungekutta) are available.
95 # Simulation time step and span can be set here.
96 hh.set_integration("euler")
97 hh.set_time_discretization(unit_list["ms"], "0.01")
98 hh.set_simulation_time_span(unit_list["ms"], "300")
99
100 # -------------------- Definitions of modules, ports, and children modules
101 list_module_name = []   # module-name list. Modules in this list will be
                               created.
102 lpq_output = {}          # "module_name:PQ" dictionary
103 llpqs_in = {}            # "module_name:PQ_list" dictionary
104 llpqs_ex = {}            # "module_name:external-defined-PQ_list" dictionary
105 llchild = {}             # "capsule-name:parent-module_name:child-module-id"
                               dictionary
106 lencapsulate_name = []   # encapsulate modules in this list
107
```

```
108 # =========================== V module
109 module_name = "Membrane potential"
110 list_module_name.append(module_name)
111 lpq_output.update({module_name:"V"})
112 llpqs_in[module_name] = ["V"]
113 llpqs_in[module_name].append("C_m")
114 llpqs_ex[module_name] = ["I_app"]
115 llpqs_ex[module_name].append("I_Na")
116 llpqs_ex[module_name].append("I_K")
117 llpqs_ex[module_name].append("I_L")
118 llchild[module_name] = ["Sodium current"]
119 llchild[module_name].append("Potassium current")
120 llchild[module_name].append("Leak current")
121
122 # =========================== I_Na module
123 module_name = "Sodium current"
124 list_module_name.append(module_name)
125 lpq_output.update({module_name:"I_Na"})
126 llpqs_in[module_name] = ["I_Na"]
127 llpqs_in[module_name].append("g_Na")
128 llpqs_in[module_name].append("E_Na")
129 llpqs_ex[module_name] = ["V"]
130 llpqs_ex[module_name].append("m_infinity")
131 llpqs_ex[module_name].append("h")
132 llchild[module_name] = ["m_infinity"]
133 llchild[module_name].append("Gate variable h")
134
135 # =========================== m_infinity module
136 module_name = "m_infinity"
137 list_module_name.append(module_name)
138 lpq_output.update({module_name:"m_infinity"})
139 llpqs_in[module_name] = ["m_infinity"]
140 llpqs_in[module_name].append("alpha_m")
141 llpqs_in[module_name].append("beta_m")
142 llpqs_ex[module_name] = ["V"]
143 llchild[module_name] = []
144
145 # =========================== h module
146 module_name = "Gate variable h"
147 list_module_name.append(module_name)
148 lpq_output.update({module_name:"h"})
149 llpqs_in[module_name] = ["h"]
150 llpqs_in[module_name].append("phi")
151 llpqs_in[module_name].append("alpha_h")
152 llpqs_in[module_name].append("beta_h")
153 llpqs_ex[module_name] = ["V"]
154 llchild[module_name] = []
155
156 # =========================== I_K module
157 module_name = "Potassium current"
158 list_module_name.append(module_name)
159 lpq_output.update({module_name:"I_K"})
160 llpqs_in[module_name] = ["I_K"]
161 llpqs_in[module_name].append("g_K")
162 llpqs_in[module_name].append("E_K")
163 llpqs_ex[module_name] = ["V"]
164 llpqs_ex[module_name].append("n")
165 llchild[module_name] = ["Gate variable n"]
166
167 # =========================== n module
168 module_name = "Gate variable n"
169 list_module_name.append(module_name)
170 lpq_output.update({module_name:"n"})
171 llpqs_in[module_name] = ["n"]
172 llpqs_in[module_name].append("phi")
173 llpqs_in[module_name].append("alpha_n")
174 llpqs_in[module_name].append("beta_n")
```

```
175 llpqs_ex[module_name] = ["V"]
176 llchild[module_name] = []
177
178 # =========================== I_L module
179 module_name = "Leak current"
180 list_module_name.append(module_name)
181 lpq_output.update({module_name:"I_L"})
182 llpqs_in[module_name] = ["I_L"]
183 llpqs_in[module_name].append("g_L")
184 llpqs_in[module_name].append("E_L")
185 llpqs_ex[module_name] = ["V"]
186 llchild[module_name] = []
187
188 # ------------------------------------- Definitions of physical-quantities
189 lpq_description = {}   # "pq_name:description" list
190 lpq_init = {}          # "pq_name:initial_value" list
191 lpq_def = {}           # "pq-name:definition" list
192 lpq_type = {}          # "pq-name:pq-type" list
193 lpq_def_type = {}      # "pq-name:definition-type" list
194 lpq_unit = {}          # "pq-name:unit" list
195
196 # =========================== state
197 pq_type = "state"
198 def_type = "ode"
199
200 name = "V"
201 lpq_description[name] = "Membrane potential"
202 lpq_init[name] = "V = -60" # add uniform(-10, 10) later
203 lpq_def[name] = "diff(V, time) = (-I_Na-I_K-I_L+I_app)/C_m"
204 lpq_type[name] = pq_type
205 lpq_def_type[name] = def_type
206 lpq_unit[name] = unit_list["mV"]
207
208 name = "h"
209 lpq_description[name] = "Proportion of inactivation h-gate at the open
                            state"
210 lpq_init[name] = "h = 0.781247"
211 lpq_def[name] = "diff(h, time) = phi*(alpha_h*(1-h)-beta_h*h)"
212 lpq_type[name] = pq_type
213 lpq_def_type[name] = def_type
214 lpq_unit[name] = "0"
215
216 name = "n"
217 lpq_description[name] = "Proportion of n-gate at the open state"
218 lpq_init[name] = "n = 0.088959"
219 lpq_def[name] = "diff(n, time) = phi*(alpha_n*(1-n)-beta_n*n)"
220 lpq_type[name] = pq_type
221 lpq_def_type[name] = def_type
222 lpq_unit[name] = "0"
223
224 # =========================== variable-parameter
225 pq_type = "variable-parameter"
226 def_type = "ae"
227
228 name = "I_L"
229 lpq_description[name] = "Leak current"
230 lpq_def[name] = "%s = g_L*(V-E_L)" % name
231 lpq_type[name] = pq_type
232 lpq_def_type[name] = def_type
233 lpq_unit[name] = unit_list["uA/cm2"]
234
235 name = "I_Na"
236 lpq_description[name] = "Sodium current"
237 lpq_def[name] = "%s = g_Na*m_infinity*m_infinity*m_infinity*h*(V-E_Na)"
                            % name
238 lpq_type[name] = pq_type
239 lpq_def_type[name] = def_type
```

```
240 lpq_unit[name] = unit_list["uA/cm2"]
241
242 name = "m_infinity"
243 lpq_description[name] = "Proportion of activation m-gate at the open state
                            (steady-state)"
244 lpq_def[name] = "%s = alpha_m/(alpha_m+beta_m)" % name
245 lpq_type[name] = pq_type
246 lpq_def_type[name] = def_type
247 lpq_unit[name] = "0"
248
249 name = "alpha_m"
250 lpq_description[name] = "Transition rate from close to open of the m-gate"
251 lpq_def[name] = "%s = -0.1*(V+35)/(exp(-0.1*(V+35))-1)" % name
252 lpq_type[name] = pq_type
253 lpq_def_type[name] = def_type
254 lpq_unit[name] = unit_list["/ms"]
255
256 name = "beta_m"
257 lpq_description[name] = "Transition rate from open to close of the m-gate"
258 lpq_def[name] = "%s = 4*exp(-(V+60)/18)" % name
259 lpq_type[name] = pq_type
260 lpq_def_type[name] = def_type
261 lpq_unit[name] = unit_list["/ms"]
262
263 name = "alpha_h"
264 lpq_description[name] = "Transition rate from close to open of the h-gate"
265 lpq_def[name] = "%s = 0.07*exp(-(V+58)/20)" % name
266 lpq_type[name] = pq_type
267 lpq_def_type[name] = def_type
268 lpq_unit[name] = unit_list["/ms"]
269
270 name = "beta_h"
271 lpq_description[name] = "Transition rate from open to close of the h-gate"
272 lpq_def[name] = "%s = 1/(exp(-0.1*(V+28))+1)" % name
273 lpq_type[name] = pq_type
274 lpq_def_type[name] = def_type
275 lpq_unit[name] = unit_list["/ms"]
276
277 name = "I_K"
278 lpq_description[name] = "Potassium current"
279 lpq_def[name] = "%s = g_K*n*n*n*n*(V-E_K)" % name
280 lpq_type[name] = pq_type
281 lpq_def_type[name] = def_type
282 lpq_unit[name] = unit_list["uA/cm2"]
283
284 name = "alpha_n"
285 lpq_description[name] = "Transition rate from close to open of the n-gate"
286 lpq_def[name] = "%s = -0.01*(V+34)/(exp(-0.1*(V+34))-1)" % name
287 lpq_type[name] = pq_type
288 lpq_def_type[name] = def_type
289 lpq_unit[name] = unit_list["/ms"]
290
291 name = "beta_n"
292 lpq_description[name] = "Transition rate from open to close of the n-gate"
293 lpq_def[name] = "%s = 0.125*exp(-(V+44)/80)" % name
294 lpq_type[name] = pq_type
295 lpq_def_type[name] = def_type
296 lpq_unit[name] = unit_list["/ms"]
297
298 # =========================== static-parameter
299 pq_type = "static-parameter"
300 def_type = "ae"
301
302 name = "I_app"
303 lpq_description[name] = "Injected current"
304 lpq_unit[name] = unit_list["uA/cm2"]
305 lpq_def[name] = "%s = 1" % name
```

```
306 lpq_type[name] = pq_type
307 lpq_def_type[name] = def_type
308 lpq_unit[name] = unit_list["uA/cm2"]
309
310 name = "phi"
311 lpq_description[name] = ""
312 lpq_def[name] = "%s = 5" % name
313 lpq_type[name] = pq_type
314 lpq_def_type[name] = def_type
315 lpq_unit[name] = "0"
316
317 name = "C_m"
318 lpq_description[name] = "Membrane capacity"
319 lpq_def[name] = "%s = 1.0" % name
320 lpq_type[name] = pq_type
321 lpq_def_type[name] = def_type
322 lpq_unit[name] = unit_list["uA/cm2"]
323
324 name = "g_L"
325 lpq_description[name] = "Leakage conductance"
326 lpq_def[name] = "%s = 0.1" % name
327 lpq_type[name] = pq_type
328 lpq_def_type[name] = def_type
329 lpq_unit[name] = unit_list["mS/cm2"]
330
331 name = "E_L"
332 lpq_description[name] = "Equilibrium potential for leak current"
333 lpq_def[name] = "%s = -65" % name
334 lpq_type[name] = pq_type
335 lpq_def_type[name] = def_type
336 lpq_unit[name] = unit_list["mV"]
337
338 name = "g_Na"
339 lpq_description[name] = "Maximum sodium conductance"
340 lpq_def[name] = "%s = 35" % name
341 lpq_type[name] = pq_type
342 lpq_def_type[name] = def_type
343 lpq_unit[name] = unit_list["mS/cm2"]
344
345 name = "E_Na"
346 lpq_description[name] = "Equilibrium potential for the sodium ions"
347 lpq_def[name] = "%s = 55" % name
348 lpq_type[name] = pq_type
349 lpq_def_type[name] = def_type
350 lpq_unit[name] = unit_list["mV"]
351
352 name = "g_K"
353 lpq_description[name] = "Maximum sodium potassium conductance"
354 lpq_def[name] = "%s = 9" % name
355 lpq_type[name] = pq_type
356 lpq_def_type[name] = def_type
357 lpq_unit[name] = unit_list["mS/cm2"]
358
359 name = "E_K"
360 lpq_description[name] = "Equilibrium potential for the potassium ions"
361 lpq_def[name] = "%s = -90" % name
362 lpq_type[name] = pq_type
363 lpq_def_type[name] = def_type
364 lpq_unit[name] = unit_list["mV"]
365
366 name = "theta_syn"
367 lpq_description[name] = ""
368 lpq_def[name] = "%s = 0" % name
369 lpq_type[name] = pq_type
370 lpq_def_type[name] = def_type
371 lpq_unit[name] = unit_list["mV"]
372
```

```
373 # ------------------------------------------------ Create modules and ports
374 id_module = {}              # dictionary of module_name:module-id
375 id_port_in = {}            # dictionary of in-port-name:in-port-id
376 id_port_out = {}           # dictionary of out-port-name:out-port-id
377 id_capsule_port_in = {}    # dictionary of capsule in-port-id
378 id_capsule_port_out = {}   # dictionary of capsule out-port-id
379 # ----------------------------------------------------
380 for module_name in list_module_name:
381
382     # Create a module using the module name
383     # "mid" is the ID of the newly created module
384     mid = mh.create(module_name)
385
386     # Store the module ID in a list of module IDs for the sake of later reuse
387     id_module[module_name] = mid
388
389     # Create physical-quantities (PQs)
390
391     # Externally-defined PQs
392     # i.e. PQs that receives values from outside of this module via in-ports.
393     for pq_name in llpqs_ex[module_name]:
394       # Create an in-port to receive the value from other module.
395       # The in-port has the same name with the physical-quantity.
396       # The port ID of the newly created is stored in a list.
397       id_port_in[mid + pq_name] = ph.create(id_module[module_name], "in",
                pq_name)
398
399       # Create a PQ
400       # pq_id is the ID of the newly created physical-quantity.
401       pq_id = pqh.create_with_type(mid, 'variable-parameter', pq_name)
402
403       # Setup the PQ
404
405       # Define description of the PQ
406       pqh.set_description(mid, pq_id, lpq_description[pq_name])
407
408       # Set unit of the PQ
409       pqh.set_unit(mid, pq_id, lpq_unit[pq_name])
410
411       # Make an association between the PQ and the corresponding in-port.
412       dh.set_definition(mid, pq_id, "0", id_port_in[mid+pq_name], "assign",
                "port")
413
414     # Internally-defined PQs
415     for pq_name in llpqs_in[module_name]:
416       # Create a PQ
417       pq_id = pqh.create(mid, pq_name)
418
419       # Setup the PQ
420       pqh.set_description(mid, pq_id, lpq_description[pq_name])
421       pqh.set_unit(mid, pq_id, lpq_unit[pq_name])
422       pqh.set_type(mid, pq_id, lpq_type[pq_name])
423
424       # The mathematical definition of the PQ is given by MathML
425       dh.set_definition(mid, pq_id, "0", \
426           cstr2mml.cstr2mml(lpq_def[pq_name]), lpq_def_type[pq_name], "")
427
428       # When the type is "state", initial value of the PQ must be defined
429       if lpq_type[pq_name] == "state":
430         dih.set_definition(mid, pq_id, "0",
                cstr2mml.cstr2mml(lpq_init[pq_name]), "ae")
431
432       # When the name of the PQ is defined as pq_output,
433       # the PQ is considered as the output of this module.
434       # Hence an out-port and association between the out-port and PQ is
                created.
```

```
435     # The out-port has the same name with the physical-quantity.
436     if pq_name == lpq_output[module_name]:
437       # Create an out-port
438       id_port_out[mid + pq_name] = ph.create(mid, "out", pq_name)
439
440       # Make an association between the out-port and the PQ
441       ph.set_reference(mid, id_port_out[mid + pq_name], pq_id)
442
443 # --------------------------------------------------------------- Make edges
444 # Structural edges.
445 # Using the information of children modules defined on a module,
446 # we can link structural edges.
447 for parent_name in llchild.keys():
448   for child_name in llchild[parent_name]:
449     bl.link_modules(id_module[child_name], id_module[parent_name])
450
451 # Functional edges.
452 # In-ports associated to physical-quantities listed in llpqs_ex require
    to be
453 # connected with functional-edges.
454 # We assume that corresponding in-port and out-port have the same name.
455 # For example, if in-port is "V" asking a value of membrane potential, the
    out-port
456 # who is a sender of the information must have the same name "V".
457 rpqs_output = dict([(p,m) for (m,p) in lpq_output.iteritems()])
458 for module_name in list_module_name:
459
460   # In-ports and associated PQs requiring functional edges to receive
      values.
461   for pq_name in llpqs_ex[module_name]:
462
463     # Find the source module and its out-port from the name of the
        out-port.
464     if pq_name in lpq_output.values():
465       source_module_name = rpqs_output[pq_name]
466       bl.link_ports(id_module[source_module_name], \
467           id_port_out[id_module[source_module_name]+pq_name], \
              id_module[module_name], \
468           id_port_in[id_module[module_name]+pq_name], "functional")
469
470 # ----------------------------- Encapsulate the membrane potential module
471 # Encapsulate
472 capsule_id = bl.encapsulate(id_module["Membrane potential"])
473 membrane_mid = id_module["Membrane potential"]
474
475 # Set name to the capsule module
476 mh.set_name(capsule_id, single_name)
477
478 # Set ports on the capsule module
479 # An in-port
480 port_name = "I_app"
481 id_capsule_port_in[capsule_id + port_name] = ph.create(capsule_id, "in",
    port_name)
482
483 # Link a forwarding edge between the in-port of the capsule module
484 # and the corresponding port of the membrane module.
485 bl.link_ports(capsule_id, id_capsule_port_in[capsule_id + port_name], \
486     membrane_mid, id_port_in[membrane_mid + port_name], "forwarding")
487
488 # An out-port
489 port_name = "V"
490 id_capsule_port_out[capsule_id + port_name] = ph.create(capsule_id, "out",
    port_name)
491
492 # Link a forwarding edge between the out-port of the capsule module
493 # and the corresponding port of the membrane module.
```

```
494 bl.link_ports(membrane_mid, id_port_out[membrane_mid + port_name], \
495     capsule_id, id_capsule_port_out[capsule_id + port_name], "forwarding")
496
497 # ---------------------- Create and link stimulus current module for test
498 module_name = "Stimulus current"
499 pq_name = "I_app"
500
501 # Create a module
502 stim_mid = mh.create(module_name)
503
504 # Create an out-port
505 id_port_out[stim_mid + pq_name] = ph.create(stim_mid, "out", pq_name)
506
507 # Create a physical-quantity
508 pq_id = pqh.create(stim_mid, pq_name)
509
510 # Setup the physical-quantity
511 pqh.set_description(stim_mid, pq_id, lpq_description[pq_name])
512 pqh.set_type(stim_mid, pq_id, lpq_type[pq_name])
513 pqh.set_unit(stim_mid, pq_id, lpq_unit[pq_name])
514 dh.set_definition(stim_mid, pq_id, "0", \
515     cstr2mml.cstr2mml(lpq_def[pq_name]), lpq_def_type[pq_name], "")
516
517 # Setup the out-port
518 ph.set_reference(stim_mid, id_port_out[stim_mid + pq_name], pq_id)
519
520 # Link a functional edge from the stimulus module to the neuron capsule
        model
521 bl.link_ports(stim_mid, id_port_out[stim_mid + pq_name], \
522     capsule_id, id_capsule_port_in[capsule_id + pq_name], "functional")
523
524 # Encapsulate the stimulus module
525 bl.encapsulate(stim_mid)
526
527 # --------------------------------------------- Save the model into file
528 file_name = current_dir + "Wang_Buzsaki_1996_interneuron.isml"
529 print "# Save model into %s" % file_name
530 save_file(file_name, format_xml(model.dump_document()))
```

Utilizing the single interneuron created by the above script, let us make an neural network model composed of N interneurons although the network model is also available in ModelDB with the name Wang_Buzsaki_1996_hippocampal_interneuron_network_30.isml. The interneurons in the network model are connected by chemical synapses with all to all connections. We need to add modules representing the synapses in the network model. An interneuron module receives a synaptic current from a synapse module. When there are N interneurons, an interneuron module receives N synaptic currents. Hence we also need to modify the single interneuron module by adding in-ports to receive synaptic currents and by re-defining the ODE of the membrane potential V to take the synaptic currents into account.

The following is the whole Python script to create an interneuron network model. This script loads the single interneuron model which has been created by the above script. The model file of the single interneuron must be in the current directory where the script is located for creating the network. Again the first 17 lines are for standalone execution of the script, and should be commented out for being called from the *insilico* terminal.

```
 1 # ---------------------------------------------------------------- Initialize
 2 # Settings for standalone executable Python script
 3 # random is necessary to set a new initial value of V
 4 import sys,os,random
 5
 6 # Absolute path to the insilico package
 7 iside_path="/.../insilicoIDE-1.x"
 8 # Current directory is used for specifying the file name to save the model
 9 current_dir=os.getcwd() + "/"
10 sys.path.append(iside_path)
11 os.chdir(iside_path)
12
13 from initialize import *
14 mlog_set_level(logging.INFO)
15 cstr2mml.set_ns_prefix("m")
16
17 model = Model(uuid4_str(), isdb)
18
19 # --------------------------------------------------- Create model handlers
20 hh = model.get_header_handler()
21 mh = model.get_module_handler()
22 ph = model.get_port_handler()
23 eh = model.get_edge_handler()
24 pqh = model.get_physicalquantity_handler()
25 dh = model.get_definition_implementation_handler()
26 dih = model.get_definition_initialvalue_handler()
27 hh = model.get_header_handler()
28 aih = model.get_article_info_handler()
29 cih = model.get_creator_info_handler()
30 uh = model.get_unit_handler()
31 bl = model.get_builder()
32
33 # ------------------------------- Load Wang_Buzsaki_1996_interneuron.isml
34 # Place Wang_Buzsaki_1996_interneuron.isml at the same directory as this
   script 35 model.load_new_isml_model(current_dir +
   'Wang_Buzsaki_1996_interneuron.isml')
36
37 # Number of modules in the network (must >= 2)
38 N = 300
39
40 # ------------------------------------------------ Get user defined units
41 unit_list = {} # "unit-code:unit-id" dictionary
42
43 ## milli-second
44 code = "ms"
45 unit_list[code] = uh.get_id_with_name("milli-second")
46
47 ## per_milli-second
48 code = "/ms"
49 unit_list[code] = uh.get_id_with_name("per_milli-second")
50
51 ## volt
52 code ="V"
53 unit_list[code] = uh.get_id_with_name("volt")
54
55 ## milli-volt
56 code = "mV"
57 unit_list[code] = uh.get_id_with_name("milli-volt")
58
59 ## simens
60 code = "S"
61 unit_list[code] = uh.get_id_with_name("simens")
62
63 ## milli_siemens_per_centi_meter2
64 code = "mS/cm2"
65 unit_list[code] = uh.get_id_with_name("milli_siemens_per_centi_meter2")
66
67 ## micro_ampere_per_centi_meter2
```

```
68 code = "uA/cm2"
69 unit_list[code] = uh.get_id_with_name("micro_ampere_per_centi_meter2")
70
71 # ------------------------------------------------------------ Header settings
72 # Single neuron module name
73 single_interneuron_name = "Wang_Buzsaki_1996_interneuron"
74
75 # Set the model name
76 hh.set_name("Wang_Buzsaki_1996_network_model")
77
78 # Settings for numerical integration
79 hh.set_simulation_time_span(unit_list["ms"], "2000")
80
81 # -------------------- Definitions of modules, ports, and children modules
82 list_capsule_in = {}    # "capsule_name:output_port_name" dictionary
83 list_capsule_out = {}   # "capsule_name:input_port_name" dictionary
84 list_module_name = []   # list of module-name.
85                         #   Modules in this list will be created or edited
86 llpqs_in = {}           # "module_name:PQ_list" dictionary
87 llpqs_ex = {}           # "module_name:external-defined-PQ_list" dictionary
88 llchild = {}            # "capsule-name:parent-module_name:child-module-id"
                             dictionary
89
90 # =========================== Single interneuron module
91 module_name = single_interneuron_name
92 list_capsule_in[module_name] = []
93 for i in range(1, N):
94    list_capsule_in[module_name].append("s_%s" % i)
95 list_capsule_out[module_name] = ["s"]
96 llchild[module_name] = ["Gate variable s"]
97
98 # =========================== V module
99 module_name = "Membrane potential"
100 list_module_name.append(module_name)
101 llpqs_in[module_name] = ["V"]
102 llpqs_ex[module_name] = ["I_syn"]
103 llchild[module_name] = ["Synaptic current"]
104
105 # =========================== I_syn module
106 module_name = "Synaptic current"
107 list_module_name.append(module_name)
108 llpqs_in[module_name] = ["I_syn"]
109 llpqs_in[module_name].append("g_syn")
110 llpqs_in[module_name].append("E_syn")
111 llpqs_ex[module_name] = ["V"]
112 for i in range(1, N):
113    llpqs_ex[module_name].append("s_%s" % i)
114 llchild[module_name] = []
115
116 # =========================== s module
117 module_name = "Gate variable s"
118 list_module_name.append(module_name)
119 llpqs_in[module_name] = ["s"]
120 llpqs_in[module_name].append("alpha")
121 llpqs_in[module_name].append("beta")
122 llpqs_in[module_name].append("theta_syn")
123 llpqs_in[module_name].append("F")
124 llpqs_ex[module_name] = ["V"]
125 llchild[module_name] = []
126
127 # ----------------------------------- Definitions of physical-quantities
128 lpq_description = {}   # "pq_name:description" list
129 lpq_init = {}          # "pq_name:initial_value" list
130 lpq_def = {}           # "pq-name:definition" list
131 lpq_type = {}          # "pq-name:pq-type" list
132 lpq_def_type = {}      # "pq-name:definition-type" list
133 lpq_unit = {}          # "pq-name:unit" list
```

```
134
135 # =========================== state
136 pq_type = "state"
137 def_type = "ode"
138
139 name = "V"
140 lpq_description[name] = "Membrane potential"
141 lpq_init[name] = "V = -60"
142 lpq_def[name] = "diff(V, time) = (-I_Na-I_K-I_L-I_syn+I_app)/C_m"
143 lpq_type[name] = pq_type
144 lpq_def_type[name] = def_type
145 lpq_unit[name] = unit_list["mV"]
146
147 name = "s"
148 lpq_description[name] = "Fraction of open synaptic ion channel"
149 lpq_init[name] = "s = 0.5"
150 lpq_def[name] = "diff(s, time) = alpha*F*(1-s) - beta*s"
151 lpq_type[name] = pq_type
152 lpq_def_type[name] = def_type
153 lpq_unit[name] = "0"
154 for i in range (1, N):
155   lpq_description[name+"_%s" % i] = lpq_description[name]
156   lpq_unit[name+"_%s" % i] = "0"   # dimensionless
157
158 # =========================== variable-parameter
159 pq_type = "variable-parameter"
160 def_type = "ae"
161
162 name = "I_syn"
163 lpq_description[name] = "Synaptic current"
164 sum_of_s = "s_1"
165 for i in range(2, N):
166   sum_of_s += "+ s_%s" % i
167 lpq_def[name] = "%s = g_syn*(V-E_syn)*(%s)" % (name, sum_of_s)
168 lpq_type[name] = pq_type
169 lpq_def_type[name] = def_type
170 lpq_unit[name] = unit_list["uA/cm2"]
171
172 name = "F"
173 lpq_description[name] = "Post-synaptic transmitter-receptor complex"
174 lpq_def[name] = "%s = 1/(1+exp(-(V-theta_syn)/2))" % name
175 lpq_type[name] = pq_type
176 lpq_def_type[name] = def_type
177 lpq_unit[name] = "0"
178
179 # =========================== static-parameter
180 pq_type = "static-parameter"
181 def_type = "ae"
182
183 name = "theta_syn"
184 lpq_description[name] = ""
185 lpq_def[name] = "%s = 0" % name
186 lpq_type[name] = pq_type
187 lpq_def_type[name] = def_type
188 lpq_unit[name] = unit_list["mV"]
189
190 name = "g_syn"
191 lpq_description[name] = "Maximal synaptic conductance"
192 lpq_def[name] = "%s = 0.001" % name
193 lpq_type[name] = pq_type
194 lpq_def_type[name] = def_type
195 lpq_unit[name] = unit_list["mS/cm2"]
196
197 name = "E_syn"
198 lpq_description[name] = "Equilibrium potential for synaptic current"
199 lpq_def[name] = "%s = -75" % name
200 lpq_type[name] = pq_type
```

```
201 lpq_def_type[name] = def_type
202 lpq_unit[name] = unit_list["mV"]
203
204 name = "alpha"
205 lpq_description[name] = "Transition rate from close to open of synaptic ion
                            channel"
206 lpq_def[name] = "%s = 12" % name
207 lpq_type[name] = pq_type
208 lpq_def_type[name] = def_type
209 lpq_unit[name] = unit_list["/ms"]
210
211 name = "beta"
212 lpq_description[name] = "Transition rate from open to close of synaptic ion
                            channel"
213 lpq_def[name] = "%s = 0.1" % name
214 lpq_type[name] = pq_type
215 lpq_def_type[name] = def_type
216 lpq_unit[name] = unit_list["/ms"]
217
218 # --------------------------------- Create and editing modules and ports
219 id_module = {}              # dictionary of module_name:module-id
220 id_port = {}                # dictionary of port-name:port-id
221 id_capsule_port_in = {}     # dictionary of capsule in-port-id
222 id_capsule_port_out = {}    # dictionary of capsule out-port-id
223 lports_sender = {}
224 lports_receiver = {}
225
226 # Initialize sender list
227 # This proccess is necessary to create out-ports in functional-unit modules
228 for list_port_name in llpqs_ex.values()+list_capsule_out.values():
229   for port_name in list_port_name:
230     lports_sender.update({port_name:None})
231
232 # Iteration on functional-unit module to create or edit
233 print list_module_name
234 for module_name in list_module_name:
235
236   if module_name == "Membrane potential":
237     # Modify the definition of the membrane potential module
238     # Get its module ID
239     mid = mh.get_ids_with_name(module_name)[0]
240   else:
241     # Otherwise create a module
242     mid = mh.create(module_name)
243
244   # Store the module ID in a list of module IDs for the sake of later reuse
245   id_module[module_name] = mid
246
247   # Create physical-quantities (PQs)
248
249   # Externally-defined PQs
250   # i.e. PQs that receives values from outside of this module via in-ports.
251   # Initialize receiver list
252   lports_receiver[module_name] = []
253   for pq_name in llpqs_ex[module_name]:
254
255     # Create an in-port to receive the value from other module.
256     # The in-port has the same name with the physical-quantity.
257     # The port ID of the newly created is stored in a list.
258     id_port[mid + pq_name] = ph.create(id_module[module_name], "in",
                                pq_name)
259
260     # Create a physical-quantity if the module type is "functional-unit"
261     if mh.get_type(mid) == "functional-unit":
262
263       # Create a PQ
264       # pq_id is the ID of the newly created physical-quantity.
```

```
265        pq_id = pqh.create(mid, pq_name)
266
267        # Setup the PQ
268        # Make an association between the PQ and the corresponding in-port.
269        pqh.set_description(mid, pq_id, lpq_description[pq_name])
270        pqh.set_type(mid, pq_id, "variable-parameter")
271        pqh.set_unit(mid, pq_id, lpq_unit[pq_name])
272        dh.set_definition(mid, pq_id, "0", id_port[mid+pq_name], "assign",
           "port")
273
274        # Update receiver list
275        lports_receiver[module_name].append(pq_name)
276
277   # Internally-defined PQs
278   for pq_name in llpqs_in[module_name]:
279
280        # Check pq_name PQ exists or not
281        if not pq_name in pqh.get_names(mid):
282          # Create a PQ
283          pq_id = pqh.create(mid, pq_name)
284        else:
285          # Get PQ ID
286          pq_id = pqh.get_id_with_name(mid, pq_name)
287
288        # Setup the PQ
289        pqh.set_description(mid, pq_id, lpq_description[pq_name])
290        pqh.set_unit(mid, pq_id, lpq_unit[pq_name])
291        pqh.set_type(mid, pq_id, lpq_type[pq_name])
292
293        dh.set_definition(mid, pq_id, "0", \
294            cstr2mml.cstr2mml(lpq_def[pq_name]), lpq_def_type[pq_name], "")
295
296        # When the type is "state", initial value of the PQ must be defined
297        if lpq_type[pq_name] == "state":
298          dih.set_definition(mid, pq_id, "0",
           cstr2mml.cstr2mml(lpq_init[pq_name]), "ae")
299
300        # When the name of the PQ is defined as pq_output,
301        # the PQ is considered as the output of this module.
302        # Hence an out-port and association between the out-port and PQ is
           created.
303        # The out-port has the same name with the physical-quantity.
304        if pq_name in lports_sender:
305
306          # Check pq_name PQ exists or not
307          if not pq_name in ph.get_names(mid, "out"):
308            # Create an out-port
309            id_port[mid + pq_name] = ph.create(mid, "out", pq_name)
310          else:
311            # Get out-port ID
312            id_port[mid + pq_name] = ph.get_ids_with_name(mid, pq_name)[0]
313
314          # Make an association between the out-port and the PQ
315          ph.set_reference(mid, id_port[mid + pq_name], pq_id)
316
317          # Update sender list
318          lports_sender.update({pq_name:mid})
319
320 # --------------------------------- Editing capsule module and its ports
321 llid_cport_s = {}          # dictionary of capsule-module-id:
                               "s_*" port-id list
322
323 # In this script, edit single_interneuron_name module
324 module_name = single_interneuron_name
325
326 # Get and store module ID whose name is module_name
327 mid = mh.get_ids_with_name(module_name)[0]
```

```
328 id_module[module_name] = mid
329
330 # Editing out-port
331 # Initialize receiver list
332 lports_receiver[module_name] = []
333 for port_name in list_capsule_out[module_name]:
334
335   # Create an out-port and store port ID
336   port_id = ph.create(mid, "out", port_name)
337   id_port[mid+port_name] = port_id
338
339   # Update receiver list
340   lports_receiver[module_name].append(port_name)
341
342 # Editing input-port and store s_* port ID
343 llid_cport_s[mid] = []
344 for  port_name in list_capsule_in[module_name]:
345
346   # Create an in-port and store port ID
347   port_id = ph.create(mid, "in", port_name)
348   id_port[mid + port_name] = port_id
349
350   # Update sender list
351   lports_sender.update({port_name:mid})
352
353   # Store s_* port ID for networking
354   llid_cport_s[mid].append(port_id)
355
356 # -------------------------------------------------------------- Make edges
357 # Structural edges.
358 # Using the information of children modules defined on a module,
359 # we can link structural edges.
360 for parent_name in llchild.keys():
361   for child_name in llchild[parent_name]:
362     bl.link_modules(id_module[child_name], id_module[parent_name],
        "structural")
363
364 # Functional or forwarding edges.
365 # Ports in lports_receiver list require to be connected with
366 # functional-edge of forwarding edge
367
368 # In-ports associated to physical-quantities listed in llpqs_ex require
    to be
369 # connected with functional-edges.
370 # We assume that corresponding in-port and out-port have the same name.
371 # For example, if in-port is "V" asking a value of membrane potential, the
    out-port
372 # who is a sender of the information must have the same name "V".
373 # In-ports and associated PQs requiring functional edges to receive values.
374 while lports_receiver:
375
376   # Get (module_name, list_port_name) tuple form lports_receiver
377   (module_name, list_port_name) = lports_receiver.popitem()
378   for port_name in list_port_name:
379
380     # Get sender IDs
381     sender_mid = lports_sender[port_name]
382     sender_port_id = id_port[sender_mid + port_name]
383
384     # Get receiver IDs
385     receiver_mid = id_module[module_name]
386     receiver_port_id = id_port[receiver_mid + port_name]
387
388     # Set edge_type
389     if ph.get_direction(sender_mid, sender_port_id) == \
390          ph.get_direction(receiver_mid, receiver_port_id):
391       edge_type = "forwarding"
```

```
392      else:
393        edge_type = "functional"
394
395      # link edge between ports
396      bl.link_ports(sender_mid, sender_port_id, \
397          receiver_mid, receiver_port_id, edge_type)
398
399
400 # ------------------------------------------------------------ Networking
401 lid_capsule = []          # list of capsule-module-id
402
403 # ========================== Modify initial value of V in "Membrane
                                  potential" module
404 mV_id = id_module["Membrane potential"]
405 pqV_id = pqh.get_id_with_name(mV_id, "V")
406
407 # Set seed
408 random.seed(1)
409 equation = lpq_init["V"] + "+ %s" % random.uniform(-10, 10)
410
411 dih.set_definition(mV_id, pqV_id, '0', cstr2mml.cstr2mml(equation), 'ae')
412
413 # ========================== Copy interneuron model N times
414 base_mid = id_module[single_interneuron_name]
415 lid_capsule.append(base_mid)
416
417 for edge_id in eh.get_ids_connected_with_head(base_mid, "any",
    "functional") \
418          + eh.get_ids_connected_with_head(base_mid, "any", "functional"):
419
420 eh.remove(edge_id)
421
422 for i in range(2, N+1):
423
424    # Copy single neuron model
425    mid = bl.copy_subtree(base_mid, "root")[0]
426    lid_capsule.append(mid)
427
428    # Set new single interneuron model name
429    mh.set_name(mid, single_interneuron_name + "_%s" % i)
430
431    # Copy s_* port ID list
432    llid_cport_s[mid] = llid_cport_s[base_mid][:]
433
434    # Modify initial value of V in "Membrane potential" module again
435    random.seed(i)
436    equation = lpq_init["V"] + "+ %s" % random.uniform(-10, 10)
437    mV_id = id_module["Membrane potential"]
438    pqV_id = pqh.get_id_with_name(mV_id, "V")
439    dih.set_definition(mV_id, pqV_id, '0', cstr2mml.cstr2mml(equation), 'ae')
440
441 # ========================== Make network
442 for head in lid_capsule:
443    for tail in lid_capsule:
444       if head != tail:
445          bl.link_ports(head, id_port[base_mid+"s"], \
446               tail, llid_cport_s[tail].pop(), "functional")
447
448 # ========================== Encapsulate Network
449 network_id = bl.encapsulate(lid_capsule)
450 mh.set_name(network_id, "Wang_Buzsaki_1996_network")
451
452 # input ports
453 port_name = "I_app"
454 network_port_id = ph.create(network_id, "in", port_name)
455
456 # Link forwarding edge
```

```
457 for tail in lid_capsule:
458   bl.link_ports(network_id, network_port_id, \
459       tail, ph.get_ids_with_name(tail, port_name)[0], "forwarding")
460
461 # ========================== Link stimulus current
462 # input ports
463 module_name = "capsule_of_Stimulus current"
464 port_name = "I_app"
465
466 mid = mh.get_ids_with_name(module_name)[0]
467 port_id = ph.get_ids(mid, "out")[0]
468
469 bl.link_ports(mid, port_id, network_id, network_port_id, "functional")
470
471 # ----------------------------------------------- Save the model into file
472 file_name = current_dir + "Wang_Buzsaki_1996_network.isml"
473 print "# Save model into %s" % file_name
474 save_file(file_name, format_xml(model.dump_document()))
```

These scripts might be seen as a conventional programming to create a mathematical models of physiological functions without using *insilico* platform. But it is not. By working in the framework of *insilico* platform, users can focus on the structure and logics of a model, and do not need to take care about algorithms for numerical calculations, of which ISSim is in charge. ISSim also can perform parallel computations for simulations of ISML models. If users want to use parallel computations on a multi-core or PC-cluster environment, usually users are required to learn techniques additionally, which is time consuming task. Moreover users can reuse other models in the script as we did for making the interneuronal network model above. Remember that the principal concept of the *insilico* platform is sharing and reuse of the existing models.

4.5   Other Model Exchange Formats

Various kinds of physiological modeling packages in variety levels such as subcellular, cellular, organic and so on have been used. This diversity of tools brings several problems, including difficulties in a reuse of models among packages and a lack of mechanisms for publishing models in electronic form. There are a couple of marked pioneering efforts addressing this problems, which have been developing versatile languages to describe mathematical models of physiological phenomena such as Systems Biology Markup Language (SBML) and CellML. Based on modular structure of ISML, we can import and slot other models written in other languages into ISML models. In this section, instead of writing a model only in ISML, we will consider to utilize other models described by other languages in *insilico* framework.

4.5.1   From Cell to Intracellular Signaling

In the previous sections we looked at the cell level dynamics, such as membrane potential and ionic currents and neuronal cell interactions. We can consider the cell level physiological phenomena as a macroscopic relative to subcellular dynamics. Since ISML is designed to represent hierarchical structure of the physiological phenomena, it is possible to describe a model that includes integration of cellular level and subcellular level phenomena. In this section, instead of modeling subcellular phenomena by ISML, we utilize models described by SBML on *insilico* framework.

The approach to understand phenomena that may happen in cells such as gene expressions and protein translations from the system-level is so called "systems biology" in which system's structures and dynamics are important targets to investigate (Kitano 2002). There is a pioneering model description language for systems biology called SBML, a free, open, XML-based format for representing biochemical reaction networks (Hucka et al. 2003; Finney and Hucka 2003). SBML is a suitable language for describing models in many areas of computational biology, including subcellular signaling pathways, metabolic pathways, gene regulation, among others.

ISML is designed to represent a functional network and hierarchical structure using its modular representation of the function. Combining SBML and ISML features can extend the capability to construct models of physiological phenomena. In *insilico* platform, there is a functional capability to import or wrap a whole SBML model into a module of ISML. Then the module can represent the subcellular phenomena which is modeled by the SBML model. By linking the module to other modules by functional and structural edges, the SBML model can be embedded in an ISML module network in the senses of both structural and functional relationships. The scheme of the hybridization of SBML and ISML is illustrated in Fig. 4.34.

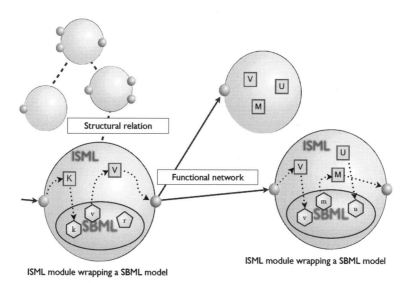

ISML module wrapping a SBML model

ISML module wrapping a SBML model

Fig. 4.34 SBML–ISML hybridization scheme. A module of ISML can import a whole SBML model in itself. By making associations between physical-quantities in the module and species/parameters in the SBML model, the SBML model is effectively imbedded in the module network. An oval in modules near bottom represents a SBML model included in the modules. Square-surrounded characters in the modules represent physical-quantities, polygon-surrounded characters in ovals are species or parameters of SBML

In SBML, there are "species" and "parameters" to represent quantitative attributes of biochemical entities. At a module including a SBML model, it is possible to define physical-quantities associated to species or parameters to set or get numerical values. The ISML part can utilize the numerical information defined in the SBML model via the physical-quantities with "get" definition that acts as a kind of one-way bridge from the SBML part to the ISML part. Similarly but with opposite direction, physical-quantities with "set" definition can affect to the SBML part from ISML part by overriding the original definition of species or parameters in the SBML model without modifying the SBML model itself. By this definition of physical-quantities, the SBML model can be functionally involved in the model.

Example of SBML–ISML Hybridization 1: Disposition of Carboxydichlorofluroscein

Let us show simple examples of the hybridization of SBML and ISML. We will begin with a very simple caricature model of the disposition of carboxydichlorofluroscein in hepatocyte (Howe et al. 2009). Although of course originally proposed model in the literature (Howe et al. 2009) includes much complicated signal transduction pathways, we extracted only two or three players among them for this example. Figure 4.35a shows a model diagram drawn by CellDesigner version 4.0.1

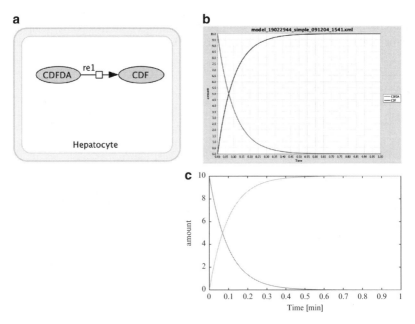

Fig. 4.35 A caricature model of the disposition of Carboxydichlorofluroscein (CDF). Carboxydichlorofluroscein diacetate (CDFD) in a cell is converted by intracellular esterases to the metabolite CDF. (**a**) A diagram of the model drawn by CellDesigner version 4.0.1. It shows only a reaction hydrolysis of CDFD to CDF. (**b**) A simulation result calculated on CellDesigner. The density of CDFDA and CDF in the cell decreases and increases, respectively. (**c**) A simulation result calculated by ISSim, which shows a consistent result with the one shown in panel (**b**)

which is developed by Systems Biology Institute devoting for editing and simulating SBML models (Funahashi et al. 2008). There is carboxydichlorofluroscein diacetate (CDFDA) in a hepatocyte which is hydrolyzed to Carboxydichlorofluroscein (CDF). The initial concentrations for CDFDA and CDF are set to 10 and $0\,\mu$M, respectively. By the hydrolytic reaction, the concentration of CDFDA decreases and contrary CDF increases as shown in Fig. 4.35b which is calculated by CellDesigner. It is worth to note that ISSim also has a capability to parse and perform simulations of SBML models. Figure 4.35c is a supplementary simulation result calculated by ISSim which showed a consistent result with the panel (b) in the same figure.

There is CDFDA in buffer (outside the cell) which can passively diffuse into the cell (Fig. 4.36a). We will expand the model by adding this component as ISML. Figure 4.36b shows a modular representation of the resultant SBML–ISML hybrid model on ISIDE. First of all, we will make a kind of SBML wrapper module. A module can import whole SBML model. On ISIDE, users can find a command "Import SBML model" in a right button context menu. Then it is possible to select a

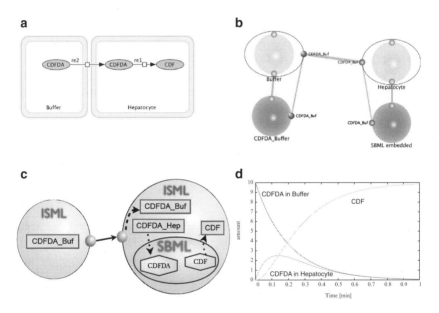

Fig. 4.36 A caricature hybrid model of the disposition of Carboxydichlorofluroscein (CDF). Carboxydichlorofluroscein diacetate (CDFDA) can passively diffuse into a cell from outside of the cell. Once CDFDA gets into a cell, it is converted by intracellular esterases to the metabolite CDF. (**a**) A diagram drawn by CellDesigner. There are two compartments, one represents the buffer and the other is a hepatocyte. Three actors are playing roles in this model, i.e. CDFDA in buffer, CDFDA and CDF in hepatocyte. (**b**) A modular representation of the model on ISIDE. There are two functional-unit modules. One represents CDFDA in buffer (outside of a hepatocyte) and the other is a kind of a wrapper module of a SBML model shown in Fig. 4.35, corresponding to a hepatocyte. (**c**) Schematic diagram of the hybridization between SBML and ISML. CDFDA in hepatocyte is overridden by a state type physical-quantity to take into account the CDFDA concentration in buffer into the SBML model. (**d**) A simulation result calculated by ISSim

SBML model file and to define associations between SBML species/parameter to physical-quantities of an ISML model. A terminal command

```
>>> module_handler.set_import_from_file(module_id, 'internal', 'sbml', \
    'cdfda_cdf_metabolism.xml')
```

will do the same task, where cdfda_cdf_metabolism.xml is the name of the SBML test model.

ISML can define two-way connections between SBML species/parameters and physical-quantities. This connection is called a *bridge* between SBML and ISML. A direction that a physical-quantity refers the value defined in a species or a parameter is called "get" and the opposite direction is "set". Once a physical-quantity gets a value in a SBML model, the value immediately becomes available in an ISML part even from other modules which are linked by functional edges. When a value of physical-quantity is "set" to a species or a parameter, the value of the species or the parameter is overridden by the value of the physical-quantity during simulations, though the SBML model is not modified explicitly. For example let us assume that there are a SBML parameter para_sbml which is set to 5, and a static-parameter type physical-quantity sp_isml which is set to 1 and is bridging to the SBML parameter with direction "set". In this case, the parameter para_sbml has the value 1 during the simulation. Let us give another example. Suppose that there are a SBML species sp_sbml which is usually a dynamical variable evolving in time according to a given ODE representing a chemical reaction, and a state type physical-quantity state_isml which has an implementation defined by an ODE and is bridging to the species with direction "set". In this case, during a simulation, the value of the species is not calculated using the given ODE in the SBML model, but is replaced by the value of the physical-quantity state_isml at every time step of the numerical integration for the simulation.

In the original small SBML model, the concentration of CDFDA in hepatocyte follows the dynamics described by

$$\frac{d}{dt}\text{CDFDA}_{\text{Hep}} = -k1 \times \text{CDFDA}_{\text{Hep}}. \tag{4.22}$$

Since now we want to involve the concentration of CDFDA in buffer, the above ODE must be modified as

$$\frac{d}{dt}\text{CDFDA}_{\text{Hep}} = -k1 \times \text{CDFDA}_{\text{Hep}} + k2 \times \text{CDFDA}_{\text{Buf}}. \tag{4.23}$$

The ODE in (4.23) is defined on a state type physical-quantity which is bridging with direction "set" to the corresponding species in the SBML model. CDFDA_Buf is defined in other module representing buffer. So we need to create an in-port to receive it (Fig. 4.36c).

We also create a variable-type physical-quantity to get the value of CDF in SBML model to monitor its dynamics. If we make an association between the physical-quantity and an out-port, the value of CDF becomes available even from the other modules (in this model we do not need to do that).

Next we create a module named CDFDA_Buffer which includes one state type physical-quantity **CDFDA_Buf** which is a concentration of the CDFDA in buffer (outside of the hepatocyte). The physical-quantity has an ODE implementation as

$$\frac{d}{dt}\text{CDFDA}_{\text{Buf}} = -k2 \times \text{CDFDA}_{\text{Buf}}. \tag{4.24}$$

Since the CDFDA passively diffuse into the hepatocyte, its concentration exponentially decreases with a velocity constant k_2.

Figure 4.36d shows a simulation result of the hybrid model calculated by ISSim, which parses the SBML part directly without converting SBML into ISML, and compiles equations for numerical computations. The concentration of CDFDA in buffer decreases monotonously and the concentration of CDF in hepatocyte, a final product, increases monotonically. CDFDA in hepatocyte once increases and then decreases.

The above example is a toy model of hybridization between a SBML model and an ISML model. However still the potential of the hybridization could be evident from the viewpoint of the multi-scale and multi-level modeling. SBML is a suitable language to describe microscopic subcellular phenomena such as signal transductions. On the other hand, ISML is originally designed to describe modularity, hierarchical structure and network of mathematical functionality of models. By combining both of them, cellular–subcellular inter level modeling can be achieved effectively. SBML models are archived, for example, in BioModels Database (www.eib.ac.uk/biomodels-main). The *insilico* platform can provide wider activity arena to users by utilizing resources in databases for SBML and ISML together to build a multi-level models.

Example of SBML–ISML Hybridization 2: Pancreatic β-Cells

Let us show another example of a SBML–ISML hybrid model briefly. Pancreatic β-cells have been drawing the attentions from experimental and computational researches because of their importance of insulin secretion (Muoio and Newgard 2008). It is known that pancreatic β-cells exhibit complex and periodic spike-burst activity in response to a rise in concentration of extracellular glucose. See Sect. 2.4 for some details. Chicago model that reproduces a membrane potential level dynamics (Fridlyand et al. 2003) includes a membrane potential, ATP/ADP concentrations, various ionic currents such as sodium, potassium, calcium, and so on. The model includes ATP-sensitive K^+ current whose channel is inhibited by high ATP concentration, resulting in membrane depolarization. This change in the membrane potential activates voltage-gated Ca^{2+} channels yielding influx of Ca^{2+}. Ca^{2+} can trigger exocytosis of insulin granules within a living organism, which is not modeled though. The model mainly focuses on the mechanism of the burst generation, in which especially Ca^{2+} slow oscillation plays an important role.

Biochemical mechanism of glucose metabolism and ATP generation by the TCA cycle within mitochondria is usually not included in such cell level models, but modeled in a scope of molecular and biochemical level studies. There is a SBML model of the glucose-stimulated insulin secretion network of pancreatic β cells (Jiang et al. 2007) which includes many entities related to glycolysis, the TCA cycle, the respiratory chain, NADH shuttles and the pyruvate cycle. This model, however, does not include membrane potential and ionic currents.

Now let us create a combined model of pancreatic β cells including membrane level dynamics conjoined with the biochemical mechanism of glucose metabolism and ATP generation. It is possible to find Chicago model written in ISML in ModelDB by searching with "chicago beta cell" in model name (Fig. 4.37a). ATP and ADP concentrations are expressed as state-type and variable-parameter type physical-quantities, respectively. We will replace these values by the ones of the latter model which is written in SBML. The SBML model can be found in EBI BioModels Database by searching with "jiang AND pancreatic AND beta AND cells" (Fig. 4.37b).

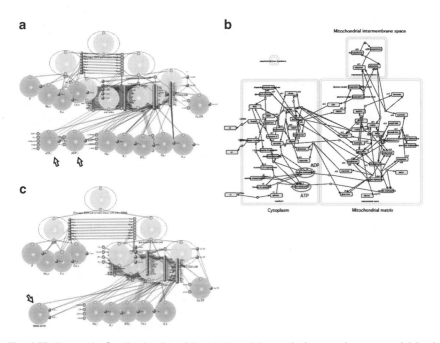

Fig. 4.37 Pancreatic β-cell related models. (**a**) A model reproducing membrane potential level dynamics, which includes ATP, ADP concentrations, several ionic currents such as voltage dependent Na$^+$, K$^+$, ATP sensitive K$^+$, Ca^{2+} and so on. The arrows indicate modules representing ATP and ADP dynamics. The diagram is drawn by ISIDE. (**b**) A biochemical model focusing of glucose metabolism and ATP generation drawn by CellDesigner. Encircled species show amount of ATP and ADP. (**c**) A hybrid model of models shown in (**a**) and (**b**). An arrow-indicated module is a wrapper module of the SBML model, which outputs ATP and ADP concentrations retrieved from the SBML model. In this model, modules representing ATP and ADP are not used. Instead of them, the values derived from the SBML model are used in other modules

Firstly we create an ISML module in the Chicago model and import the SBML model into the module. Then we define variable-parameter type physical-quantities bridging to species in the SBML model to get the values of ATP and ADP concentrations. Modules representing ATP and ADP in the Chicago model are removed, and instead of them, the module wrapping the SBML model provides the dynamics of ATP and ADP concentrations derived from the SBML model to other ISML modules. By this procedure, the SBML model is slotted in the hybrid model.

ISSim can perform simulations of SBML–ISML hybrid models. Figure 4.38 shows simulation results of the original Chicago model (the ISML model) in left panels and of the SBML–ISML hybrid model in right panels. Since there is no feedback path from the ISML part to SBML part, such as calcium density in cytoplasm affecting to mitochondrial TCA cycle for example, the dynamics of ATP amount shown in Fig. 4.38d is identical to the one in the original SBML model, which periodically fluctuates with higher frequency than the ATP concentration in the original Chicago model does (Fig. 4.38c). The difference in the fluctuation frequency causes the difference in the membrane potential behaviors as shown in panels (a) and (b) in the same figure.

This attempt to develop SBML–ISML hybrid models is still preliminary at the moment. However this approach has a potential to be an authentic and effective multi-level modeling protocol.

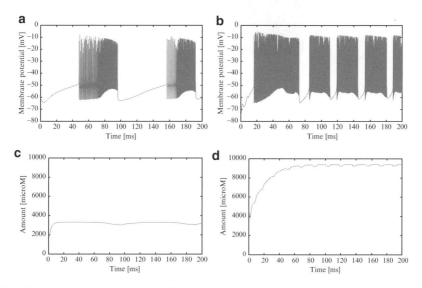

Fig. 4.38 Sample simulation results of pancreatic β-cell models. (**a, c**) Results of the Chicago model written in ISML, and (**b, d**) of the SBML–ISML hybrid model. (**a**) and (**b**) show the membrane potential exhibiting characteristic oscillatory bursts. (**c**) and (**d**) show the time course of ATP amount in cytoplasm

4.5.2   Yet Another Format

CellML[1] (Cuellar et al. 2003) is an open format based on the XML developed by
the Auckland Bioengineering Institute at the University of Auckland. The purpose
of CellML is again similar to the ones of SBML and of ISML, that is, to store, share
and reuse computer-based mathematical models. The CellML model repository on
www.cellml.org is openly available.

ISIDE can load models written in CellML format. For example, let us find a
model that reproduces the action potential of ventricular myocardial fibres proposed
by Beeler and Reuter (1977) in CellML model repository. There is a CellML model
simulator called OpenCell which also provides a simple text-based interface to edit
the model. It is available freely from their website. Figure 4.39a shows a snap shot
of OpenCell on which a simulation result of the Beeler Reuter model is displayed.
ISIDE can load a CellML model, and display it based on the modular representation
of the model on *insilico* canvas by converting CellML into ISML (Fig. 4.39b). By
this feature, it is very natural to include and integrate CellML models as building
blocks with an ISML model to create a new model. A simulation result by ISSim
is shown in Fig. 4.39c which is consistent with the result in panel (a) in the same

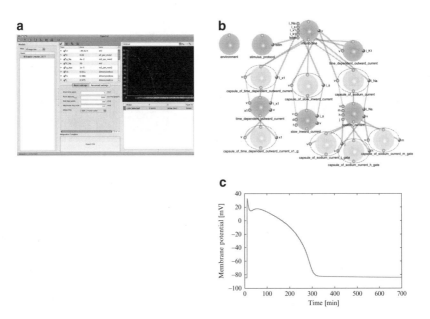

Fig. 4.39 Integration of models written in CellML format. (**a**) A snapshot of OpenCell version
0.7, a simulator for CellML models. (**b**) Modular representation of Beeler Reuter model originally
written in CellML model and loaded in ISIDE. (**c**) A simulation result of the model calculated by
ISSim

---

[1] http://www.cellml.org/.

figure. It is worth to note that ISIDE can export a model in CellML format if the ISML model includes only the elements that CellML supports. For example if an ISML model includes timeseries data or morphological data which CellML does not support, ISIDE does not export the model into CellML format.

4.6 Integration of Morphological and Timeseries Data on an ISML Model

Morphology is a basis of physiological functions. In models of physiological functions, morphological data can be used to define boundaries of a space for agent-based simulations (see Sect. 2.6), a shape of a medium on which electrical excitation propagates, a domain on which partial differential equations (PDEs) are solved, and so on. Morphological data can be defined in a module in ISML, and be referred from other modules which are structurally connected directly and non-directly to the module including the morphological data. The morphology can be described by mathematical equations such as $0 \leq x \leq 1$ and $0 \leq y \leq 1$ representing a square field, pre-defined primitive components such as circle and cuboid, or by an aggregation of spatially-distributed vertices listed in a finite element method mesh format. Such morphological data can be written either in the ISML model or in a separate file. In the latter case, only the reference path to the data file is written in the ISML model.

Morphological data defined in a module is composed of several pieces. In an example case shown in Fig. 4.40 a morphology is composed from three pieces (object 1 cuboid, and objects 2 and 3 cylinders). Each of them is needed to be specified its shape type such as cuboid, cylinders, etc and dimensions such as length of sides,

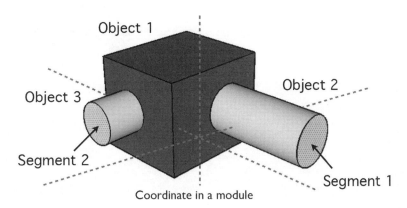

Fig. 4.40 An example of morphological data defined in a module, which is composed of three pieces, i.e. object 1 a cuboid, and objects 2 and 3 cylinders in this example. Each piece is needed to be given spatial location and rotation on the coordinate defined on the module. Segments specify particular parts in the morphology to be used for definition of boundary conditions, for example

height, etc. At the same time spatial position and rotation within the coordinate defined on the module must be given. To specify particular parts in the morphology, "segment"s can be defined and are used in, for example, definition of boundary conditions.

4.6.1   Visualization of ISML Models

The physical structure of physiological functions can be visualized based on morphological information assigned to each module, which can be helpful to users to understand the model more intuitively. Figure 4.41a shows a whole human body model (Yamasaki et al. 2003). Body segments such as trunk, lumbus and upper arm, and the joints such as hip, shoulder and knee, are modeled as modules. Spatial locations and angles of each segment are represented as physical-quantities. In this model, morphological data of bone of each body segment is also assigned to the corresponding module. The hierarchical structure is used to represent spatial and logical relationship between segments.

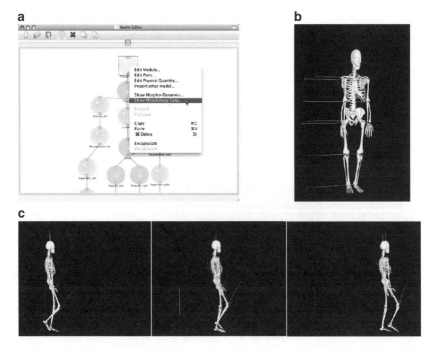

Fig. 4.41 Integration of morphological data and timeseries data in an ISML model. (**a**) A whole body model. Modules represent head, trunk, left clavicle pelvis, left elbow joint, right knee joint, etc. Each module defines its own spatial position and rotation in relative coordinate with respect to its parent module. Trunk is considered to be a origin of the body coordinate

In the model, modules representing left and right hip-joints are modeled as child modules of the module representing the pelvis. The rotation angle and relative spatial position of the hip-joints are defined as a relative coordinate system with respect to their parent module, that is, the pelvis. A module representing neck is logically considered as a child of the trunk module because the trunk acts as a kind of origin of the coordinate defined on the body. In this case, the physical spatial position of the neck is defined on the relative coordinate to the trunk. Assignment of morphological data to a module can be done with a command "Edit Morphology" in the right-click context menu on ISIDE.

The model including morphological data can be visualized based on the morphology and spatial configuration information defined in each module as shown in Fig. 4.41b. In the right-click context menu, users can find a command "Show Morphological Data" which creates a new window to show the morphology visualization. It is possible to rotate and scaling of the objects by mouse action.

Another function of ISIDE is to integrate timeseries data obtained either by numerical simulations or physiological experiments into models. Each physical-quantity can be defined an association to a set of timeseries data. This is very similar idea to point an in-port as definition of a physical-quantity, but instead of pointing an in-port ID, it points a timeseries data set. Then during a simulation, the physical-quantity associated to a timeseries data takes a value from the timeseries data at every time step of a numerical integration.

Usually one timeseries data includes several records. And each record is assigned to a physical-quantity. Hence for convenience once the data file is loaded in ISIDE by choosing a command from menu "Load timeseries data", and ID is given to each record. Then at each physical-quantity setup dialog, a set of timeseries is assigned to a physical-quantity using the timeseries ID. In order to help to make association between a physical-quantity and a set of timeseries data, an XML based markup language, referred to as TSML, has been developed within the framework of *insilico* platform.

Figure 4.41c shows an example how the morphological data and timeseries data are used to display the model. The timeseries data of each joint angle obtained from a motion capture experiment is imported to ISIDE. The motion captured data of human body movement is formatted in TSML (Time Series Markup Language) that is defined in the *insilico* platform. The timeseries data, morphological data and the mathematical expressions are merged in a model on ISIDE to visualize the human walking movement in three-dimensional space. A command "Show Morpho-Dynamics" opens a window in which an animation is played.

4.6.2 PDEs Solved on Morphological Data

To solve the PDE we used the finite element method (FEM) which requires morphological data for defining a domain on which PDEs are solved, an initial and boundary conditions, and PDEs as governing equations. ISIDE can export a model including PDEs and morphological data into the freeFEM++ format (Gfem format). FreeFEM++ is a free software of FEM solver developed at Pierre et Marie Curie University (www.freefem.org/ff++). PDEs are transformed into the weak form in the conversion process because freeFEM++ requires.

In Sect. 3.3, the simulation of the excitation propagation on a 1-dimensional domain was illustrated. Here as a simple extension of such simulation, a model and simulation of the excitation propagation on a 2-dimensional square sheet is exemplified. The model can be found in the ModelDB with the model name FitzHugh–Nagumo_monodomain_2D_model.isml. The excitable medium is modeled by the FHN model (FitzHugh 1961; Nagumo et al. 1962) in PDE form described as follows:

$$\frac{\partial}{\partial t}u = \epsilon\left(\frac{\partial^2}{\partial x^2}u + \frac{\partial^2}{\partial y^2}u\right) + \lambda\left(v - u\left(1 - u\right)\left(u - \theta\right)\right), \qquad (4.25)$$

$$\frac{\partial}{\partial t}v = \alpha u - \beta v,$$

where x and y represent spatial axes variables.

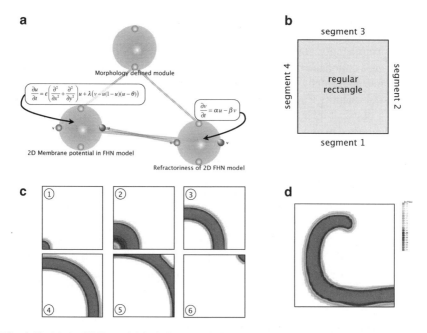

Fig. 4.42 (a) An ISML model including morphological data and PDEs. PDEs are solved on the domain defined by the morphological data. (b) The morphological data defined on the top module in (a). Four side edges of the rectangle are declared as segments 1, 2, 3 and 4 which are used to specify locations to define boundary conditions. (c) A result of FEM simulation calculated by FreeFEM++. (d) Another simulation result using the same PDEs and morphology but with a different initial condition with an additional stimulus current

In Fig. 4.42a, the top module includes a morphological information that is described as a primitive components "regular-rectangle" with side length of 1 cm (Fig. 4.42b). Modules located in a sublayer of the module having the morphological information, i.e. two modules representing the membrane and refractoriness of the FHN model, can refer and use the morphological information as domain on which PDEs are solved.

Besides morphological data, segments of morphology are described in a module to be used for defining initial and boundary conditions. In the example, as shown in Fig. 4.42b each of four side edges (outline box) of the rectangle domain is declared as one segment on which a boundary condition is set. Segments can be specified as many as necessary, and one boundary condition can be defined on each segment.

When PDEs are defined in physical-quantities, boundary and initial conditions must be also defined. In the example, the boundary conditions are given as the Neumann type, i.e. the value that derivative of a solution takes on the boundary is given, using the segments (Fig. 4.42b) as follows:

$$\frac{\partial u}{\partial y} = \frac{\partial v}{\partial y} = 0 \quad \text{on the segments 1 and 3,} \tag{4.26}$$

$$\frac{\partial u}{\partial x} = \frac{\partial v}{\partial x} = 0 \quad \text{on the segments 2 and 4.}$$

The initial conditions are

$$u_0(x, y) = 1 - \frac{1}{1 + \exp\left(\left(-50\sqrt{x^2 + y^2} + 5\right)\right)}, \tag{4.27}$$

$$v_0(x, y) = 0.$$

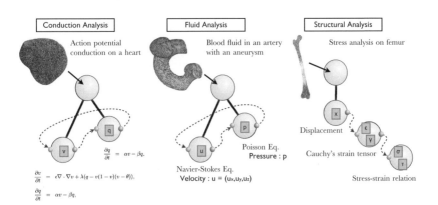

Fig. 4.43 Example problems to be solved with PDEs on morphological information. From the left, conduction analysis, fluid analysis, and structural analysis

A simulation of a model written in Gfem format which is generated by a conversion from the ISML model can be performed by FreeFEM++, and a result is shown in Fig. 4.42c. The excitation initiated at the left bottom corner of the rectangle domain propagates over the entire region. If we apply an additional stimulus at proper timing, the spiral excitation can be evoked as shown in Fig. 4.42d.

There are many varieties of problems to be considered with PDEs and morphological information in physiological phenomena. Other examples of such problems are shown in Fig. 4.43. From the left, conduction analysis which is the same type problem we considered previously (Fig. 4.42), fluid analysis, for example, simulating blood flow in an artery with an aneurysm, and structural analysis for simulating stress distribution on the femur. By using *insilico* platform, thanks to the modular representation of entities composing models, it is possible to try experimental simulations easily with, for example, keeping the morphology and replacing only governing equations, or on the contrary with different morphology and the same equations. These modifications of the model are done simply by replacing modules.

Chapter 5
Key Concepts of *insilico*ML

Mathematical models, in particular dynamic system models of physiological functions in from molecules and cells to individual organisms play a key role for integration of vast stores of knowledge on physiology since they are capable of describing time evolution of biological system states quantitatively based upon physical and chemical principles or phenomenological logic governing system behavior. The number of mathematical models of biological functions published in peer reviewed journals and complexity of each of those models rapidly increase as computational performance increases. This raises difficulties in reproducing simulated behaviors of the published models and sharing the models by third parties, hindering the promotion of sciences and knowledge integration. *Insilico*ML (ISML) has been developed to overcome this problem in parallel with other pioneering efforts to define XML languages such as SBML and CellML. In Chap. 4 the focus was on how to use ISIDE to create models written in ISML. In this chapter, specification of ISML is described in rather detail.

5.1 Overview of Model Representation in ISML

We consider a target biological system as an aggregate of elements referred to as "modules". Modules are characterized by a name, physical-quantities representing dynamic or static states with their mathematical implementations. These mathematical implementations specify the dynamics of how the states evolve in time and space, and the model geometry such as morphological model (shape), posture and position, among others. Modules can functionally affect to each other by giving and receiving values of their physical-quantities. Input and output interfaces of a module, called in-ports and out-ports respectively, to access to physical-quantities in a module are defined on a module. A physical-quantity associated to an in-port can receive and refer the inbound value to the in-port. The sender of the value is an out-port on the other module which is indicated by an "edge". The out-port sends a value of a physical-quantity associated to itself. This functional relationship can be defined between any two modules with direction specified by its departure (head)

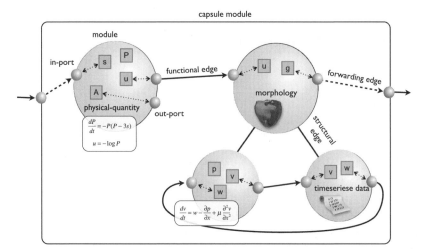

Fig. 5.1 ISML 1.0 entities to describe a model. This figure includes modules, edges, ports, physical-quantities and the encapsulation of modules. For example, the value of physical-quantity u in left upper module goes out through an out-port and is transmitted to the right module as indicated by the edge. The value is referred by physical-quantity u through an in-port and utilized within the module. In this example, all modules are encapsulated by a capsule module (outer largest one), which means that modules are tightly packed and cannot directly access to or be accessed by modules exterior to the capsule module. The paths indicated by forwarding edges linked to the ports of the capsule module are the only ways for having communication between modules inside and outside of the capsule module

and destination (tail) ports. Figure 5.1 summarizes all elements appearing in ISML modeling and relationship among them.

The edges can also represent structural and logical relationships between two modules. The edges with structural and logical (also and capsular) types allow the user to construct biophysical models with modular, hierarchical representation. Similarly edges with functional and forwarding types allows to construct network representations.

Modules can also import morphological data (numerical and analytical) and time-series data acquired by experiments and/or model simulations, and utilize them in the model. With this feature, ISML 1.0 allows the users to combine experimental and theoretical/model-based research.

Another important idea implemented in ISML 1.0 is called "capsulation". A set of modules can be tightly packed by declaring capsulation, and contained by a capsule module which is a kind of a representative of the modules within the capsule (Fig. 5.1). The capsule module can have in- and out-ports which provide the sole interfaces for access to and from the modules inside the capsule. In-ports of a capsule module are linked to in-ports of encapsulated modules by "forwarding" edges. The edge is called so because of the analogy with the concept of port-forwarding used in computer networks. Similarly, some out-ports of modules inside the capsule module are connected to out-ports of the capsule module by "forwarding" edges.

The capsulation can nest, leading to a hierarchical representation of a model, i.e. a capsule module can include other capsule modules. In many cases, a capsule module simply symbolizes a certain physiological function, and can be easily reused as an element of other models.

ISML can describe various types of mathematical models including ordinary differential equations (ODE), partial differential equations (PDE), difference equations, and agent-based simulation models that utilize IF-THEN rules among others. These models represent the dynamics of physiological functions and the geometry, i.e., morpho-metrics of living organisms underlying the functions. Morphology and timeseries data can be integrated to the models on ISIDE. Models can be numerically calculated by insilico simulator or can be converted in to C++ or JAVA codes including numerical integration algorithm (Fig. 5.2).

As described in Sect. 4.5.2, ISML 1.0 is compatible with CellML which is one of pioneering efforts to develop extensible markup language to describe mathematical models of biological functions. Let us briefly summarize the similarities between ISML and CellML. Both languages describe physical units for values used

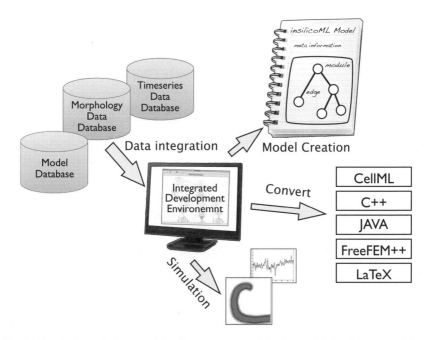

Fig. 5.2 Typical use of ISML models. ISML cooperates with other model-development and simulation environments such as model databases, databases of morphological and time series data, and other computer languages used for model simulations. ISML can be converted to other formats such as C++, CellML, xppaut ode file, and LaTeX. ISML models can be parsed, newly constructed, modified, and simulated using integrated development environments. An ISML model is composed of several modules representing functional elements (modules) of the model. The modules are linked to each other by edges which represent structural, logical and functional relationships among modules. Modules can include morphological and time series data, if necessary, to describe a model characterized by its geometry and dynamics of the modules

in models. A basic element comprising a model is called a "component" in CellML, while its correspondence is "module" in ISML 1.0. Physical variables called as "variable" in CellML can be read as "physical-quantity" in ISML 1.0. For ISML 1.0, a physical-quantity can be specified as one of several types such as "state", "dynamic-parameter", "static-parameter", and "func-expression" among others. In particular, a major difference from CellML is that the physical-quantity of ISML 1.0 has a type of "morphology" and "timeseries" to deal with numerically and analytically defined morphology and time series data acquired by real biological experiments and/or numerical simulations. The morphology type physical-quantity in ISML 1.0 is required, for example, in a problem with partial differential equations that uses morphology to define the shape of domain and boundary of the problem, or in a problem with ordinary differential equations that has constraint conditions defining a non-trivial manifold (e.g. equations of motion of a mass constrained on a surface with a complicated shape).

In- and out-ports to export/import values are specified in ISML 1.0, but there are no element having clear correspondence in CellML. In CellML, this concept is indirectly expressed as public-/private-interfaces of variables. The "connections" in CellML are used to express equivalence between different variables and components, and this roughly corresponds to the edges defined in ISML. Relationships indicated by edges and connections contribute to the construction of physiological ontology based on a large set of models. There are several differences between CellML connections and ISML edges. The edges in ISML represent directional relationships among modules whereas CellML connections are non-directional. That is, each edge in ISML possesses "head" and "tail" modules to be linked. Moreover, each edge can have an operation type "meaning" to which several words are assigned to specify how two modules are affected to each other.

5.2 Expressions Used in ISML 1.0

In this section, the term "insilico model", "ISML model", or simply "model" refers to a single model described by an entire ISML document which includes meta-information, definition of physical units, morphological data and time series data as well as mathematical expressions representing dynamics of the physical functions under consideration.

5.2.1 Modularity and Linkage Among Modules

Here we describe the Beeler–Reuter (BR) model (Beeler and Reuter 1977) which is appeared in Sect. 4.5.2 in order to illustrate the modularity and linkage among modules in ISML 1.0. The BR model can simulate electrical excitation of a cardiac ventricular cell membrane. The dynamics of the membrane potential V are determined by four ionic channel currents, i.e. time-independent potassium outward current, time-activated outward current, fast inward sodium current, and slow inward calcium current, and another current as an external stimulus. The dynamics of each ionic current is determined by the state of the corresponding ion channel and the membrane potential. For example, the slow inward calcium current i_s can be described as

$$i_s = \bar{g}_s \cdot d \cdot f \cdot (V_m - E_s) \tag{5.1}$$

where V_m is the membrane potential and E_s the reversal potential of i_s. d and f are the gate variables whose dynamics depend on the membrane potential.

Let us consider a model representing the slow calcium channel current i_s designed as in Fig. 5.3. The module I_s defining the current i_s is designed with three sub-modules which are graphically represented by the three ball-like modules connected by the solid gray lines (representing the structural edges) in Fig. 5.3. In this case, the variables d, f and E_s on the right-hand side of (5.1) are specified as physical-quantities in the modules gate_d, gate_f and E_s, respectively. To define i_s in the module I_s, these three physical-quantities are imported through the functional edges indicated by the solid curves in Fig. 5.3. V_m is obtained externally via the in-port of this model, and is used by the modules gate_d, gate_f and I_s as indicated by the dashed curves (representing the forwarding edges).

Capsule module of slow calcium channel current

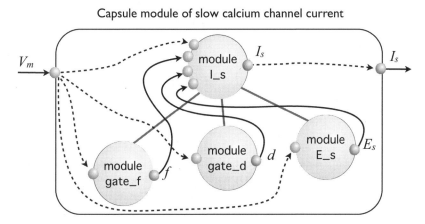

Fig. 5.3 A schema of the insilico model of the slow calcium channel current modeled in the Beeler–Reuter model. *Circles* and a *square frame* represent functional-unit modules and a capsule module, respectively. *Black solid, gray solid* and *dashed lines* represent structural, functional and forwarding edges, respectively. *Small circles* on modules represent ports (*left*: input, *right*: output). The module I_s defines the electrical current $I_s = \bar{g}_s \cdot d \cdot f \cdot (V_m - E_s)$ receiving the membrane potential V_m from the outside of the capsule module, a reversal potential E_s from a module E_s and two gate variables d and f, respectively, from the modules gate_d and gate_f. A constant conductance $\bar{g}_s$ is defined in module I_s

5.2.2 Edge

The edges shown in Fig. 5.3 are listed in an ISML document in the **edge-set** section. This section specifies the type of each edge, its departure (**head**) and destination (**tail**), and **operation** describing the meaning of its functional relationship with a verb or a verb phrase.

```
<edge-set >
<edge type="functional" edge-id="CBDFC06C-07D8-...-C30D2AE48D3A">
<head module-id="E72OOPSA-EE46-...-0FAC08E14333" port-id="3"/>
<tail module-id="A6CD8604-2F61-...-DBD7C92378E5" port-id="2"/>
<operation > gate </operation >
</edge >
<edge type="structure" edge-id="OD4JC011A-MSC3-...-CAUE2AE4334A">
<head module-id="A6CD8604-2F61-...-DBD7C92378E5" port-id="0"/>
<tail module-id="R12ZQPSA-31D6-...-BVA3LME173E3" port-id="0"/>
<operation > constitute of </operation >
</edge >
   :
   :
</edge-set >
```

The head and tail of each edge of functional and forwarding types specify both module-id and port-id to uniquely identify a port within the model. In the case of structural, logical and capsular edges, port-id=0 is used as a dummy port since these edges define the relationship between modules and do not use actual ports such as in- or out-ports.

5.2.3 Module

A description of the module I_s starts with <**module** module-id="A6CD8604-2F61-...-DBD7C92378E5" type="functional-unit"/>, where the module-id is specified as a 16 byte universally unique identifier (UUID) which is unique across all insilico models. The type of the module I_s is specified as *functional-unit*, meaning that this module possesses particular features such as **physical-quantities** necessary for modeling physiological functions.

ISML 1.0 offers several types of modules other than *functional-unit*. These include *container* and *capsule*. A module of *container* type does not possess any **physical-quantities** and is supposed to represent a conceptual box to put several modules together, e.g., a module of a *container* type with a name "ganglion" may be composed of many modules representing nerve cells. A *capsule* type module acts as a symbol of a physiological function which is modeled by a set of modules. A capsule module itself thus does not possess any **physical-quantities** but may have only in-/out-ports as its interfaces. A module has a property called "template" which is similar to the concept of a class in C++. If a module has a property "template = true", it is allowed to realize an object instantiation. Dynamic construction and destruction of instances of a template module can be managed during run-time of numerical simulations. Rules which are necessary for instance management can also be described in ISML 1.0. This is particularly important for constructing and simulating agent-base models. A **module** section is also characterized by the following four children tags: **property**, **port-set**, **physical-quantity-set** and **morphology-set**. The **property** describes basic properties of a module, such as name, keywords, track and so on. The tags **keywords** and **track** are described in the next section.

5.2.4 Port

Getting back to the module I_s example, it has four in-ports and one out-port. An ISML document enumerates them in the **port-set** section in the **module** section.

```
<port-set >
<port direction="out" port-id="1" ref-physical-quantity-id="1">
<name > I_s </name >
<description > Slow calcium channel current </description >
</port >
<port direction="in" port-id="2">
<name > V_m </name >
<description > Membrane potential </description >
</port >
    ⋮
</port-set >
```

In this example only two of the five ports in module I_s are described for brevity. Each port has a direction (in or out) and is given a port-id as a sequential number unique within a module. A port with direction="in" is called an in-port, and "out" an out-port. A physical-quantity that goes out through this out-port must be specified by its ID (physical-quantity-id) in an attribute **ref-physical-quantity-id** of the port. In a case of in-ports, ID of the associated physical-quantity is not written here, instead the ID of the in-port is specified in the definition of the physical-quantity.

5.2.5 Physical-Quantity

In the ISML paragraph for port described above, the physical-quantity pointed by the ref-physical-quantity-id="1" in out-port section corresponds to the slow calcium current which is defined as a *variable-parameter* in the **physical-quantity-set** section in the module as follows

```
<physical-quantity type="variable-parameter" physical-quantity-id="1">
<name > I_s </name >
<value-type-set >
<value-type precision="double" unit-id="12"/>
</value-type-set >
<dimension type="scalar"/>
<implementation type="ae" format="mathml"/> ... </implementation >
</physical-quantity >
```

A **type** and **physical-quantity-id** must be given first when a physical-quantity is defined. In this example, I_s is declared as the *variable-parameter*, which expresses values that vary during simulation, but which is not dynamic variable defining its derivative. The values obtained by this module are defined in the **implementation** with mathematical formulae. ISML 1.0 defines other types of physical-quantities, i.e. *state, variable-parameter, func-expression, nominal, morphology* and *timeseries*. State type is used for defining dynamic variables accompanied by differential equations. In the example above, the **value-type-set** describes its precision set to *double* (other possible candidates are int, char, and bool) and unit specifying a **unit-id** as a user-defined combination of the fundamental seven base units, (i.e., metre, kilogram, second, ampere, kelvin, candela, mole) and radian together with prefixes such as kilo, milli, and micro. The **dimension** is set to *scalar* in this example. ISML 1.0 supports vector and matrix values for a physical-quantity to use in equations.

The **implementation** section is used to describe concrete contents of the physical-quantity such as values for *variable-* and *static-parameters* and mathematical formulae for *states*. Specific definition is described in a child tag **definition**. For example, the *variable-parameter* I_s is defined as

```
<implementation >
<definition type="ae" format="mathml">
<m:math >
... Here is MathML expression for $i_s = \bar{g}_s \cdot d \cdot f \cdot (V_m - E_s)$ ...
<m:math >
</definition >
</implementation >
```

The contents described in the **definition** are categorized by specifying one of the following types: *ode* (ordinary differential equations), *pde* (partial differential equations), *dde* (delay differential equations), *sde* (stochastic differential equations), *de* (difference equations), *ae* (algebraic equations), *assign, func-expression* and *conditional*. The user can specify arithmetic, algebraic, and statistic expressions includes vectors and matrices in equations and inequalities described by MathML in the **definition** tag.

The implementation of a variable type physical-quantity V_m in the module I_s receiving the value of the membrane potential from the in-port is as follows:

```
<implementation >
<definition type="assign" sub-type="port" format="reference">
<reference port-id="2"/>
</definition >
</implementation >
```

Simply the ID of the in-port is specified. By this implementation, this physical-quantity and the in-port is associated to each other.

5.2.6 Capsulation

The module l_s and its sub-modules are encapsulated in the example above. The capsule itself is also represented as a module with the *capsule* type depicted as a square frame in Fig. 5.3. The capsule module is linked to a top module among the encapsulated modules by a capsular edge (which is the module l_s in the example and the capsular edge is not illustrated in the figure). Once the root module is encapsulated, in- and out-ports of the modules under the root module can be linked to the input and out-ports of the capsule module by *forwarding* type edges. Whether or not the module is encapsulated is specified in the **property** tag of the module as

```
<module type="functional-unit" module-id="5D2778FE-AC24-...-4474C60F4816">
<property >
<capsulation state="true">
<capsulated-by module-id="C0EA5AE5-16DD-...-8A5848ACF217"/>
</capsulation >
```

where the module-id of the capsule module is specified.

Figure 5.4 shows the whole BR model composed of the module (membrane) modeling the membrane potential and four encapsulated modules representing the ionic currents (rightmost one is the slow calcium current module shown in Fig. 5.3). The BR model is encapsulated by the capsule module named "Beeler Reuter 1977 model" which has an in-port to receive the external stimulus current and an output to export the membrane potential. Using this model the user can create, for example, coupled BR models by linking several individual BR models.

5.3 SBML–ISML Hybridization

In Sect. 4.5.1 SBML and ISML hybrid models are introduced. A module can import an entire SBML model in itself. The imported SBML model is written in **import** tag in the module section. The **import** has an attribute **type** which takes either "internal"

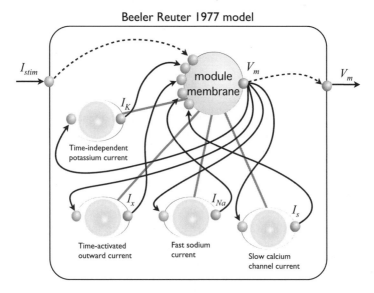

Fig. 5.4 A schema of the Beeler–Reuter model composed of a membrane and four ionic currents which are modeled using *insilico*ML 1.0 and encapsulated for further reuse. The capsule module of slow calcium current at the bottom right of "Beeler Reuter 1997 model" is the one shown in Fig. 5.3. *Circles* surrounded by circle-frame represent capsule modules

or "external", and **format** takes "sbml". When internal is specified, a whole SBML document is included in the module section. While if external is given, only reference such as a file name with path is described here.

<module type="functional-unit" module-id="5D2778FE-AC24-...-4474C60F4816">
<import type="internal" format="sbml">
<sbml level="2" version="4" xmlns="http://www.sbml.org/sbml/level2">
Here is SBML document
</sbml >
</import >

or more simply with external type,

<module type="functional-unit" module-id="5D2778FE-AC24-...-4474C60F4816">
<import type="external" format="sbml" iref="./sbml-model.xml">

where **iref** is used to specify a path in a local file system, and **xref** is used for remote filesystem.

Definition of relationships between physical-quantities and elements defined in the SBML model imported in the module is described at **bridge** section in the **implementation** of physical-quantities. A **bridge** tag has an attribute **type** which takes "sbml", which means that the imported model is written in SBML format. Another attribute **direction** taking either "set" or "get" defines a functional meaning of the relationship. **sub-type** attribute takes either "species" or "parameter", defining the sort of the target element in the imported SBML model to be associated to this physical-quantity. ISML description for bridge is as follows:

```
<module type="functional-unit" module-id="5D2778FE-AC24-...-4474C60F4816">
<implementation >
<bridge type="sbml" direction="get" sub-type="species">
<connector type="species">s_1</connector >
</bridge >
```

Connectors in a bridge defines a specific target to which the physical-quantity is associated. A **type** attribute takes either "species", "reaction" or "parameter". When sub-type of the bridge is species, a connector's type must be also "species", and the value of the **id** attribute of species tag in a listOfSpecies tag of a SBML model is given. When sub-type of the bridge is parameter, a connector with type "parameter" takes the value of the id listed in listOfParameter in the SBML model. To specify a parameter defined in a reaction, the value of the reaction id listed in listOfReaction in the SBML model is also needed to be given as a connector with type="reaction".

```
<bridge type="sbml" direction="get" sub-type="parameter">
<connector type="reaction">re_11</connector >
<connector type="parameter">p_2</connector >
</bridge >
```

When direction="get", the value defined in the target element in the SBML model is taken and assigned to the physical-quantity. In this case, the implementation of the physical-quantity is as follows:

```
<implementation >
<definition type="assign" sub-type="bridge"/>
<bridge type="sbml" direction="get" sub-type="species">
<connector type="species">s_1</connector >
</bridge >
<implementation >
```

which means that the value for this physical-quantity comes through the bridge. The type of the physical-quantity is variable-type.

On the contrary when direction="set", the value defined in the physical-quantity is sent to the element (parameter or species) in the imported SBML model, and the value of the element is replaced by the value of the physical-quantity. If the element in the imported model has a dynamics (described by a differential equation, for example), the dynamics is completely overridden by the one defined in the physical quantity.

```
<implementation >
<definition type="ode">
<m:math >
... Here is MathML expression of an ODE ...
</m:math >
</definition >
<bridge type="sbml" direction="set" sub-type="species">
<connector type="species">s_6</connector >
</bridge >
<implementation >
```

5.4   Morphological Data and Mathematical Expressions

Let us take another example, a reaction-diffusion dynamics with the FitzHugh–Nagumo model which is similar to the one considered in Sect. 4.6.2. This model represents the excitation conduction on a two dimensional medium as a model of cardiac tissue analyzed by Hall and Glass (1999). They investigated spatio-temporal changes in the propagating action potential on a two-dimensional sheet paved with excitable cell models. The reaction diffusion equations that they used are as follows:

$$\frac{\partial v}{\partial t} = \frac{1}{\epsilon}\left(v - \frac{1}{3}v^3 - w\right) + D\nabla^2 v + I_{loc} + I_{stim}(t) \tag{5.2}$$

$$\frac{\partial w}{\partial t} = \epsilon(v + \beta - \gamma w)g(v) \tag{5.3}$$

where $g(v)$ is a sigmoidal function controlling the rate of the pacemaker,

$$g(v) = \frac{w_H - w_L}{1 + exp(-kv)} + w_L. \tag{5.4}$$

v and w are the excitation and recovery variables, respectively. I_loc is a constant current applied to a localized region at the center of the sheet. $I_{stim}(t)$ is a pulsatile stimulation current used for resetting applied to selected points at selected timings.

We begin by designing the corresponding insilico model using ISML 1.0 as shown in Fig. 5.5. The PDE with FitzHugh–Nagumo is modeled here by two modules representing excitation variable v and recovery variable w, respectively, which are encapsulated. The modules M_v and M_w are mutually linked by functional edges. The module M_v receives two external electric current stimuli I_stim and I_loc via the in-ports of the capsule module. The module M_domain provides a morphology of the tissue on which the reaction diffusion problem is solved. Here the module M_domain is linked by a structural (constituent) edge to the capsule

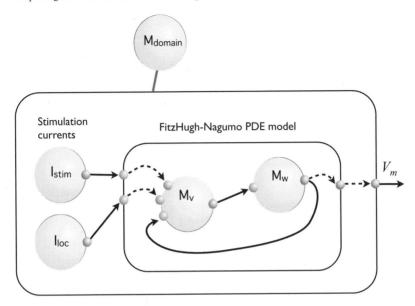

Fig. 5.5 A schema of the reaction-diffusion PDE with the FitzHugh–Nagumo model receiving two external stimulation currents. See the caption of Fig. 5.3 for representations

module including the modules representing the current stimulations as well as the PDE of the FitzHugh–Nagumo model, since all of them are solved on the morphology provided by the module M_domain.

5.4.1 Morphology

The module M_domain is implemented as a *functional-unit* type in which morphological data is described. Morphological data is described either as a list of two or three dimensional coordinates of vertices, pre-defined primitive shape with dimensions, or as analytical (mathematical) expressions. In the case of list of points, the data can be kept as an independent file either in a hard disk at a local machine or in a remote database. In other cases, type of pre-defined primitive shape or mathematical expressions are written. Morphological data is described in the **morphology** section in a module. The simple two-dimensional square sheet used in this example (shown in Fig. 5.6a) can be described using primitive shape expression as follows:

```
<morphology >
<geometry-set >
<geometry type="solid" geometry-id="1">
<layout >
<origin >
```

a 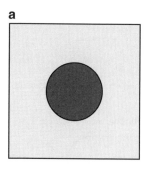   **b**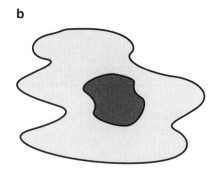

Fig. 5.6 (**a**) A simple two-dimensional square sheet (tissue) used in the reaction-diffusion system with the FitzHugh–Nagumo model. (**b**) An example of the complicated two-dimensional tissue that may be defined using numerical morphological data

```
<m:math >
... Here is MathML expression for a vector (0,0,0) ...
</m:math >
</origin >
<rotation type="eulerangle">
<unit unit-id="8"/>
<first axis="x">0.2</first >
<second axis="y">1.1</second >
<third axis="x">0</third >
</rotation >
</layout >
<constitution type="primitive" sub-type="regular-rectangle">
<length-unit unit-id="1"/>
<parameter-set >
<parameter type="side">1.0</parameter >
<parameter-set >
</constitution >
</geometry >
<geometry type="solid" geometry-id="2">
<layout >
<origin >
<m:math >
... MathML for a vector (0,0,0) ...
</m:math >
</origin >
</layout >
<constitution type="primitive" sub-type="circle">
<angle-unit unit-id="8"/>
<parameter-set >
<parameter type="radius">0.5</parameter >
```

```
<parameter-set >
</constitution >
</geometry >
</geometry-set >
</morphology >
```

Each **geometry** represents one piece that composes a whole morphology. A sequential unique ID number within the module is assigned to each fragment (geometry) defining the morphology. Each geometry is characterized by rotation angle and spatial position in a coordinate sticked to the module that are described in **layout** tag as well as shape information described in **constitution** tag. The morphology defined in the module M_domain can be more complicated, such as shown in Fig. 5.6b, defined numerically as morphological data. In that case, the description of ISML is as follows;

```
<morphology >
<geometry-set >
<geometry type="solid" geometry-id="1">
<constitution type="external" format="vtk">
<unit unit-id="8"/>
<reference iref="./data/morphology/domain.vtk"/>
</constitution >
</geometry >
</geometry-set >
</morphology >
```

Note that if the resource file defining the morphology is at a remote machine, the attribute xref is used to specify the remote address.

ISML 1.0 defines a special type of physical-quantity named *shape* which is one of ISML-predefined physical-quantities, i.e. every module possesses this physical-quantity by default when the module includes morphological data. It is a special case of the *morphology* type and it can be used for binding morphological data. The physical-quantity *shape* represents the shape of a module. The value of this physical-quantity can be referenced by any modules located in lower layer of the module. In the example (Fig. 5.5), the modules M_v, M_w, I_stim and I_loc can refer the morphological data defined in the module M_domain. For example, assume that the module M_v has a morphology type physical-quantity with ID=4. In its definition in the implementation, the physical-quantity is assigned morphological data by inheriting the physical-quantity *shape* of the parent module M_domain. This is done by specifying the module-id of the parent module as follows;

```
<implementation >
<definition type="assign" format="morphology">
<reference module-id="A6CD8604-2F61-...-DBD7C92378E5"/>
</definition >
</implementation >
```

5.4.2 Partial Differential Equation

The module M_v includes a partial differential equation as the governing equation of the dynamics of *v* which is a state type physical-quantity. For the state
type physical-quantity, unlike the variable-parameter type, **argument-set**, **domain**
and **problem-condition** must be defined. By declaring the argument variables in
argument-set, it becomes clear that the state type physical-quantity used in partial
differential equations is a function of those argument variables. For example, state
type physical-quantity *v* of the FitzHugh–Nagumo PDE model is a function of time
and space (*x* and *y*). The arguments are also used for the func-expression type
physical-quantity as shown later. The **domain** specifies the domain on which the
function or differential equations are evaluated or solved. The following is a section
of ISML 1.0 describing **argument-set** and **domain** for our PDE example:

```
<argument-set >
<argument physical-quantity-id="2">
<argument physical-quantity-id="3">
</argument-set >
<domain >
<definition format="morphology">
<reference physical-quantity-id="4"/>
</definition >
</domain >
```

In the example, we are dealing with the reaction-diffusion equations defined on
a two-dimensional sheet. The two arguments declared in this ISML fragment
correspond to the spatial coordinate variables *x* and *y* which are independently
defined elsewhere as the *nominal* type physical-quantities with ID=2 and 3. The
nominal type physical-quantity does not have its **implementation** in their definition,
but only has the specifications of **name**, **unit**, **precision** and **dimension**. The domain
is defined by specifying a morphology type physical-quantity (ID=4) assigning
morphological data from the parent module as mentioned above. The domain can
also be defined mathematically in MathML, in which the argument variables define
functions.

In the implementation of the state type physical-quantity *v* for example, the partial differential equation is described in MathML format. All terms of the equations
are written on the left hand side of the equation, and the right hand side is always
0. Only for PDE definition, the **definition** tag has a sub-type attribute to specify the
type of PDE as one of *elliptic*, *parabolic*, *hyperbolic* and *others*, and a **form** attribute
specifying if the PDE is represented as *weak* using variational formulation or *strong*
using PDE with derivatives.

```
<implementation >
<definition type="pde" sub-type="parabolic" form="strong" format="mathml">
<m:math >
```

... Here is MathML expressions for

$$\frac{\partial v}{\partial t} - D\frac{\partial^2 v}{\partial x^2} - D\frac{\partial^2 v}{\partial y^2} + \frac{1}{\epsilon}\left(v - \frac{1}{3}v^3 - w\right) - I_{loc} - I_{stim}(t) = 0 \dots$$

</m:math >

</definition >

</implementation >

In the example of the FitzHugh–Nagumo model, zero-flux boundary conditions are applied to the four borders of the square area, which is equivalent to the zero Neumann condition at each boundary. These conditions must be described in the model. In general, ISML 1.0 specifies conditions accompanying the equations to be solved, such as boundary conditions, in the **problem-condition** section. The *constraint*, *material* and *initial* conditions as well as *boundary* condition can be defined. For the definitions of the boundary conditions, the user must specify a type of either *neumann*, *dirichlet*, *robin* or *other*. If necessary (for example in the cases of force and flow), the direction can be defined in the **direction** tag by specifying argument variables such as x or y, or explicitly by a word, either *perpendicular* or *parallel*, indicating the direction with respect to the tangential plane (or line) at every point on the boundary. Notice that thus *perpendicular* is the direction normal to the boundary. The notation *Neumann*(·) is used to describe the Neumann condition. For example, if the Neumann condition of state v is 0 on a segment of the boundary (specified by a segment-id described later), it is written as *Neumann*(v) = 0 as shown in the example ISML document below:

<problem-condition-set >

<problem-condition type="boundary">

<configured-at segment-id="2"/>

<direction > perpendicular **</direction >**

<definition type="neumann" format="mathml">

<m:math >

<apply > **<eq />**

<apply > **<ci** type="function"> Neumann**</ci >**

<ci > v **</ci >**

</apply >

<cn > 0 **</cn >**

</apply >

</m:math >

</definition >

</problem-condition >

</problem-condition-set >

Similarly *Dirichlet*(·) is for defining the Dirichlet condition. The user can define other type conditions using general mathematical formulae.

For constraint conditions, a space (usually a manifold) M_x that satisfies $x \in M_x$ for x being the target *physical-quantity* to be constrained is defined in the definition section. M_x can be described either in a mathematical format such as $\{x \mid f(x) = 0\}$

with format="mathml" or a set of points represented by morphological data for digitized, numerically defined cases with format="morphology". To specify material conditions, the values of physical material properties such as Young's modulus, density, and viscosity are set in an equation format.

For those conditions to be complete, it is necessary to specify spatial positions where each of those conditions is applied. Particular segments constituting the morphology, i.e., subsets of points of numerical morphological data or pieces of geometric objects such as an arc or a portion of rectangle, are defined in **segment-set** section in the definition of the **morphology**, and are labeled by a sequential ID (segment-id) to be specified uniquely within the module. Each segment is also categorized either by *point*, *line*, *area* or by *volume*. Then an appropriate condition is applied to each segment specified by its segment-id in the **configured-at** tag. The following is an example of the definition of the segments.

```
<morphology >
<segment-set >
<segment segment-id="1" type="line">
<piece id="1"/>
<piece id="3"/>
<piece id="5"/>
</segment >
</segment-set >
</morphology >
```

In the **problem-condition**, conditions such as *mesh*, *solver*, and *post* can be defined. They are related to numerical methods and processing performed during and after simulations of the model. The *mesh* condition specifies the number of nodal points involved in the specified structure, and it is used by a mesh-generator. Mesh-node information is described in a separate file with a certain format. Note that instead of specifying the number of nodes in the structure, it is also possible to import external node files. The *solver* condition specifies a solver used for the finite element method to solve PDE with morphology. ISML 1.0 offers at least six choices; LU, Cholesky, Crout, CG, GMRES and UMFPACK. Convergence criteria must be specified. The *post* condition is used to define the post-processors. The post-processor type determines the file format to which simulation results are saved, such as gnuplot and medit.

In our example of the FitzHugh–Nagumo model, the module I_loc generates the constant current applied to a localized region at the center of the sheet (indicated as a gray disk in Fig. 5.6a, b) in order to either make the media in the corresponding region to oscillate or to depress excitability. The current can be described mathematically as a function of spatial coordinates x and y: $I_{loc}(x, y) = C_{low}$ when $x^2 + y^2 \leq r^2$, otherwise $I_{loc}(x, y) = C_{high}$, where C_{low} and C_{high} are constants. Thus the physical-quantity transmitted from I_loc to the module M_v is not a scalar value but is a set of values as the function of the space. To describe this sort of physical-quantity, the *func-expression* type physical-quantity is available. The *func-expression* type physical-quantity does not provide values but a set of mathematical formulae of the function as it is (as text descriptions). The *func-expression* type physical-quantity also requires specification of the argument variables for use in the definition of the **implementation** of the *func-expression*. The definition of the *func-expression* type physical-quantity begins with <**physical-quantity** type="func-expression" physical-quantity-id="1">. It is characterized by **name**, **precision, dimension, unit, argument, domain** and **implementation**. Note that to define the **dimension**, in this case, the row and/or col must be specified as *discretization-dependent* (i.e., <**dimension** type="matrix"> <**col** >discretization-dependent</**col**> <**row** >discretization-dependent</**row**> </**dimension**>), since the dimension of such a variable cannot be specified until a solver discretizes the domain of the function defined in the *func-expression* for numerical computations. Example code of the **implementation** for the module I_loc is as follows:

```
<implementation >
<definition type="conditional">
<case-set case-set-id="1">
<case substructure-id="1">
<condition format="mathml">
<m:math >
... Here is MathML expression for x² + y² ≤ 4 ...
</m:math >
</condition >
<definition type="func-expression" format="mathml">
<m:math >
... Here is MathML expression for I_loc(x, y) = 0 ...
</m:math >
</definition >
</case >
<case substructure-id="2">
<definition type="func-expression" format="mathml">
<m:math >
... Here is MathML expression for I_loc(x, y) = 10 ...
</m:math >
</definition >
</case >
</case-set >
</definition >
</implementation >
```

In the **implementation**, the current I_loc, which is the function of spatial coordinates x and y, is defined using conditional branches, since the value of I_loc depends on the position, for example, IF $x^2 + y^2 \leq r^2$ THEN $I_{loc} = C_{low}$, ELSE $I_{loc} = C_{high}$. The conditional branches are described by the **definition** tag with type="conditional" which takes **case-set** tags as children. There are several **cases** in a **case-set**. The condition is described in MathML format and the definition is evaluated if the condition is satisfied. These are described in the **case** section labeled by **substructure-id**. The **case**s in the same **case-set** are considered to be in the same context, which is similar to the "IF – ELSE IF – ELSE" syntax in computer languages. The last **case** in the example code above, which has no **condition** section, corresponds to the "ELSE" phrase. The type of the definition in a **case** can be again "conditional" among other types such as "ode" and "ae". By this specification, nesting IF phrases, such as

```
IF condition A: then
    IF condition B: then
        Definition 1
    ELSE:
        Definition 2
ELSE:
    Definition 3
```

can be implemented. The attribute **substructural-id** is used to specify a **definition** uniquely in a **implementation** section.

5.5 Databasing, Tracking, and Relating Models

Models created by the user are stored in a model database (ISML database) if the models satisfy certain criteria, such as publication in peer reviewed journals. Models in the database are uniquely identified by a database ID (db-id), which is assigned by the database server when the model is registered in it. Users can download models from the database and reuse them to develop new models. ISML 1.0 is capable of tracking any module in terms of the history of its reuse if the module is registered in the database. This is done using the **property** section, where an ISML document of each module records the db-id of the models that involve (reuse) the module. The reuse history of modules can provide another basis for the model-based ontology of physiological functions as well as the meta-information assigned to the modules and the edges of **operation** type with "meaning" as mentioned above.

Let us take the BR model again as an example for illustrating how the reuse history is recorded in each module. When creating the BR model, we first create the model of the slow inward calcium channel current (Fig. 5.3) as an element of the BR model. During development of the calcium channel model until it is completed and registered in the ISML model database, the model ISML document does not have a database ID (db-id). In this period, when a module such as the module I_s is created in the calcium channel model, module tracking information such as "module I_s is involved in this calcium channel model" is written as a property of the module I_s as follows:

```
<track >
<involved db-id="this" date="2007-06-06"/>
</track >
```

where "this" corresponds to the calcium channel model. This indicates that the module l_s is used as an element of the calcium channel model. When the development of the calcium channel model is completed and it is uploaded to the ISML database, the database server assigns a unique db-id to the calcium channel model. More precisely, the db-id is given to the capsule module encapsulating the model of the calcium channel. Let us assume that the server has given database ID 41 to the calcium current model. Then the value "this" in the **track** tag for the module l_s is replaced by the newly assigned db-id at the server, and the tracking record is rewritten as <**involved** db-id="41" data="2008-06-06"/>.

In the subsequent development process of the entire BR model (Fig. 5.4), let us assume that the slow calcium current model has already been registered in the database, and we can download it from the database for reuse as one of the modules composing the BR model. When the calcium channel model is reused in the BR model, a new **track** record is added to the module l_s which has been a sub-module of the calcium channel model as follows:

```
<track >
<involved db-id="this" date="2008-03-22"/>
<involved db-id="41" date="2007-06-06"/>
</track >
```

In this way, the reuse history of the module l_s is accumulated and logged.

The Luo–Rudy (LR) model (Luo and Rudy 1991) is another representative model describing the action potential generation of the ventricular cardiac myocyte. The LR model can be considered as an extended or modified version of the BR model. Luo and Rudy modified the model by adding several new modules including representations of extracellular ion concentrations, a novel potassium channel and a plateau potassium current. The slow calcium current model was used in the LR model the same way as formulated in the BR model. Let us consider ISML model creation according to this actual history of the model development. That is, we reuse the slow calcium channel model of the BR model for the creation of the LR model using ISML 1.0. Upon the registration of the newly created LR model, tracking for module l_s is revised as;

```
<track >
<involved db-id="131" date="2008-08-14"/>
<involved db-id="75" date="2008-03-22"/>
<involved db-id="41" date="2007-06-06"/>
</track >
```

where ISML models with bd-id $= 41$, 75 and 131 correspond to the slow calcium channel model, the BR model, and the LR model, respectively. Once the module is created and the model including that module is registered, this history track is maintained as long as the module exists even if the module is modified (deletion or edit of some part of the module). Because of this, modifications made to the module cannot be identified clearly from the **track** records, although one can try to compare the module after repeated modifications and the original module which is kept unchanged in the database. Nevertheless, because it is expected that the essential nature of the module will remain even after modification, it is worthwhile to provide the history tracking capability.

Based on the tracking information recorded in each module, the origin of each module involved in a model can be revealed, and, with appropriate analytical tools for ISML documents on the database, the user can obtain a tree diagram of the module (model) development for analysis of module phylogeny. In our simple example, the user can discover that the BR model and the LR model use the same or similar slow calcium channel model. Moreover, the track information enables the user to define distance or similarity between two different models. The larger the number of common modules shared in the two models, the smaller the distance between the models, leading to a model-based construction of the physiological ontology.

References

Ait-Haddou R, Kurachi Y, Nomura T (2010) On calcium buffer dynamics within the excess buffer regime. J Theor Biol 264:55–65

Armstrong CM, Bezanilla F (1977) Inactivation of the sodium channel. II. Gating current experiments. J Gen Physiol 70:567–590

Asai Y, Nomura T, Sato S, Tamaki A, Matsuo Y, Mizukura I, Abe K (2003) A coupled oscillator model of disordered interlimb coordination in patients with parkinson's disease. Biol Cybern 88:152–162

Asai Y, Suzuki Y, Kido Y, Oka H, Heien E, Nakanishi M, Urai T, Hagihara K, Kurachi Y, Nomura T (2008) Specifications of insilicoml 1.0: a multilevel biophysical model description language. J Physiol Sci 58(7):447–458

Bassingthwaighte JB (2000) Strategies for the physiome project. Ann Biomed Eng 28:1043–1058

Beeler GW, Reuter H (1977) Reconstruction of the action potential of ventricular myocardial fibres. J Physiol 268(1):177–210

Bertram R, Butte MJ, Kiemel T, Sherman A (1995) Topological and phenomenological classification of bursting oscillations. Bull Math Biol 57(3):413–439

Cuellar AA, Lloyd CM, Nielsen PF, Bullivant DP, Nickerson DP, Hunter PJ (2003) An overview of cellml 1.1, a biological model description language. Simulation 79:740–747. http://www.cellml.org/

Doi S, Sato S (1995) The global bifurcation structure of the BVP neuronal model driven by periodic pulse trains. Math Biosci 125(2):229–250

Faber GM, Silva J, Livshitz L, Rudy Y (2007) Kinetic properties of the cardiac L-type Ca^{2+} channel and its role in myocyte electrophysiology: a theoretical investigation. Biophys J 92(5):1522–1543

Fabiato A (1985) Time and calcium dependence of activation and inactivation of calcium-induced release of calcium from the sarcoplasmic reticulum of a skinned canine cardiac Purkinje cell. J Gen Physiol 85:247–289

Finney AM, Hucka M (2003) Systems biology markup language: level 2 and beyond. Biochem Soc Trans 32:1472–1473

FitzHugh R (1961) Impulses and physiological states in theoretical models of nerve membrane. Biophys J 1:445–466

Fridlyand LE, Tamarina N, Philipson LH (2003) Modeling of ca2+ flux in pancreatic β-cells: role of the plasma membrane and intracellular stores. Am J Physiol Endocrinol Metab 285(1 48-1): E138–E154, cited By (since 1996):35

Funahashi A, Matsuoka Y, Jouraku A, Morohashi M, Kikuchi N, Kitano H (2008) Celldesigner 3.5: a versatile modeling tool for biochemical networks. Proc IEEE 96(8):1254–1265

Glass L, Josephson ME (1995) Resetting and annihilation of reentrant abnormally rapid heartbeat. Phys Rev Lett 75(10):2059–2062

Hall K, Glass L (1999) How to tell a target from a spiral: the two probe problem. Phys Rev Lett 82(25):5164–5167

Heien E, Asai Y, Nomura T, Hagihara K (2009) Optimization techniques for parallel biophysical simulations generated by insilicoide. IPSJ Trans Adv Comput Syst J 2(2):131–143

Hill AV (1938) The heat of shortening and the dynamic constants of muscle. Proc R Soc B Lond 126:136–195

Hille B (2001) Ion channels of excitable membranes, 3rd edn. Sinauer Associates, Sunderland

Hodgkin AL, Huxley AF (1952) A quantitative description of membrane current and its application to conduction and excitation in nerve. J Physiol 117(4):500–544

Howe K, Gibson GG, Coleman T, Plant N (2009) In silico and in vitro modeling of hepatocyte drug transport processes: importance of abcc2 expression levels in the disposition of carboxy-dichlorofluroscein. Drug Metab Dispos 37(2):391–399

Hucka M, Finney A, Sauro H, Bolouri H, Doyle J, Kitano H, Arkin A, Bornstein B, Bray D, Cornish-Bowden A, Cuellar A, Dronov S, Gilles E, Ginkel M, Gor V, Goryanin I, Hedley W, Hodgman T, Hofmeyr J, Hunter P, Juty N, Kasberger J, Kremling A, Kummer U, Le Novre N, Loew L, Lucio D, Mendes P, Minch E, Mjolsness E, Nakayama Y, Nelson M, Nielsen P, Sakurada T, Schaff J, Shapiro B, Shimizu T, Spence H, Stelling J, Takahashi K, Tomita M, Wagner J, Wang J (2003) The systems biology markup language (sbml): a medium for representation and exchange of biochemical network models. Bioinformatics 19(4):524–531. http://sbml.org/

Hunter PJ, Borg TK (2003) Integration from proteins to organs: the physiome project. Nat Rev Mol Cell Biol 4:237–243

Izhikevich EM (2006) Dynamical systems in neuroscience: the geometry of excitability and bursting. The MIT Press, Cambridge, MA

Jiang N, Cox RD, Hancock JM (2007) A kinetic core model of the glucose-stimulated insulin secretion network of pancreatic β-cells. Mamm Genome 18(6–7):508–520

Kawazu T, Nakanishi M, Suzuki Y, Nomura T (2007) A platform for in silico modeling of physiological systems. In: 29th Annual international IEEE EMBS conference proceedings, pp 1394–1397

Keener J, Sneyd J (2009) Mathematical physiology I: cellular physiology, 2nd edn. Springer, New York

Keizer J, Smolen P (1991) Bursting electrical activity in pancreatic beta cells caused by $Ca^{(2+)}$- and voltage-inactivated Ca^{2+} channels. Proc Natl Acad Sci USA 88(9):3897–3901

Kitano H (2002) Systems biology: a brief overview. Science 295:1662–1664

Kubo R (1966) The fluctuation-dissipation theorem. Rep Prog Phys 29:255–284

Lee EH, Hsin J, Mayans O, Schulten K (2007) Secondary and tertiary structure elasticity of titin Z1Z2 and a titin chain model. Biophys J 93(5):1719–1735

Li H, Linke WA, Oberhauser AF, Carrion-Vazquez M, Kerkvliet JG, Lu H, Marszalek PE, Fernandez JM (2002) Reverse engineering of the giant muscle protein titin. Nature 418(6901):998–1002

Luo CH, Rudy Y (1991) A model of the ventricular cardiac action potential. Depolarization, repolarization, and their interaction. Circ Res 68(6):1501–1526

Maruyama K, Kimura S, Yoshida H, Sawada H, Kikuchi M (1984) Molecular size and shape of β-connectin, and elastic protein of striated muscle. J Biochem (Tokyo) 95:1423–1493

McMahon TA (1984) Muscles, reflexes, and locomotion. Princeton University Press, Princeton, NJ

Muoio DM, Newgard CB (2008) Mechanisms of disease: molecular and metabolic mechanisms of insulin resistance and β-cell failure in type 2 diabetes. Nat Rev Mol Cell Biol 9(3):193–205, cited By (since 1996):102

Nagumo J, Arimoto S, Yoshizawa S (1962) An active pulse transmission line simulating nerve axon. Proc IRE 50:2061–2070

Noble D (1962) A modification of the Hodgkin–Huxley equations applicable to Purkinje fibre action and pacemaker potentials. J Physiol 160:317–352

Nomura T, Glass L (1996) Entrainment and termination of reentrant wave propagation in a periodically stimulated ring of excitable media. Phys Rev E 53(6):6353–6360

Nomura T, Sato S, Doi S, Segundo JP, Stiber MD (1994) Global bifurcation structure of a Bonhoeffer–van der Pol oscillator driven by periodic pulse trains. Biol Cybern 72:55–67

Panfilov AV, Holden AV (1997) Computational biology of the heart. Wiley, Chichester

Reimann P (2002) Brownian motors: noisy transport far from equilibrium, Phys. Rep. 361, pp. 57–265.

Rinzel J (1977) Repetitive nerve impulse propagation: numerical results and methods. In: Fitzgibbon WE, Walker HF (eds) Research notes in mathematics – nonlinear diffusion. Pitman, London

Rinzel J, Lee YS (1987) Dissection of a model for neuronal parabolic bursting. J Math Biol 25(6):653–675

Roden DM (2004) Drug-induced prolongation of the QT interval. N Engl J Med 350(1):1013–1022

Rudy Y, Silva JR (2006) Computational biology inthe studyof cardiac ion channels and cell elec-trophysiology. Q Rev Biophys 39(1):57–116

Rybak IA, Shevtsova NA, Lafreniere-Roula M, McCrea DA (2006) Modelling spinal circuitry involved in locomotor pattern generation: insights from deletions during fictive locomotion. J Physiol 577:617–639

Sanguinettia MC, Mitchesonb JS (2005) Predicting drug – hERG channel interactions that cause acquired long QT syndrome. Trends Pharmacol Sci 26(3):119–124

Segundo JP, Altshuler E, Stiber M, Garfinkel A (1991) Periodic inhibition of living pacemaker neu-rons (I): locked, intermittent, messy, and hopping behaviors. Int J Bifurcat Chaos 1(3):549–581

Sherman A, Bertram R (2005) Integrative modeling of the pancreatic β-cell. http://www.math.fsu.edu/~bertram/papers/beta/encyclopedia.pdf

Suzuki Y, Asai Y, Kawazu T, Nakanishi M, Taniguchi Y, Heien E, Hagihara K, Kurachi Y, Nomura T (2008) A platform for in silico modeling of physiological systems II. Cellml compatibility and other extended capabilities. In: 30th Annual international IEEE EMBS conference proceedings, pp 573–576

Suzuki Y, Asai Y, Oka H, Heien E, Urai T, Okamoto T, Yumikura Y, Tominaga K, Kido Y, Nakanishi M, Hagihara K, Kurachi Y, Nomura T (2009) A platform for in silico modeling of physiological systems III. In: 31th Annual international IEEE EMBS conference proceedings, pp 2803–2806

Tadross MR, Dick IE, Yue DT (2008) Mechanism of local and global Ca^{2+} sensing by calmodulin in complex with a Ca^{2+} channel. Cell 133:1228–1240

Tanskanen AJ, Greenstein JL, Chen A, Sun SX, Winslow RL (2007) Protein geometry and place-ment in the cardiac dyad influence macroscopic properties of calcium-induced calcium release. Biophys J 92:3379–3396

Wang X, Buzsáki G (1996) Gamma oscillation by synaptic inhibition in a hippocampal interneu-ronal network model. J Neurosci 16(20):6402–6413

Watanabe H, Sugiura S, Kafuku H, Hisada T (2004) Mutiphysics simulation of left ventricular fill-ing dynamics using fluid-structure interaction finite element method. Biophys J 87:2074–2085

Yamasaki T, Nomura T, Sato S (2003) Possible functional roles of phase resetting during walking. Biol Cybern 88(6):468–496

Yoshino K, Nomura T, Pakdaman K, Sato S (1999) Synthetic analysis of periodically stimulated excitable and oscillatory membrane models. Phys Rev E 59(1):956–969

Index